# Business Informa

| Company Name: | |
|---|---|
| Address: | |
| | |
| | |
| | |
| Phone: | Fax: |
| Email: | |
| Log Book No.: | |
| Log Started: | |
| Log Ended: | |
| File Location: | |
| Notes: | |

| **Title:** | | |
|---|---|---|
| **Duration:** | **Meeting Date:** | **Meeting Time:** |
| **Location:** | **Called By:** | **Type:** |
| **Facilitator:** | **Note Taker:** | **Time Keeper:** |
| **Attendees:** | | |

## Agenda

| **Topic:** | **Presenter:** |
|---|---|
| 1. | |
| 2. | |
| 3. | |
| 4. | |
| 5. | |

| | Action | Person Responsible | Deadline | |
|---|---|---|---|---|
| 1. | | | | ❐ |
| 2. | | | | ❐ |
| 3. | | | | ❐ |
| 4. | | | | ❐ |
| 5. | | | | ❐ |
| 6. | | | | ❐ |
| 7. | | | | ❐ |
| 8. | | | | ❐ |
| 9. | | | | ❐ |
| 10. | | | | ❐ |

**Notes:**

# Agenda

| Topic: | Presenter: |
|---|---|
| 1. | |
| 2. | |
| 3. | |
| 4. | |
| 5. | |

| | Action | Person Responsible | Deadline | |
|---|---|---|---|---|
| 1. | | | | ❐ |
| 2. | | | | ❐ |
| 3. | | | | ❐ |
| 4. | | | | ❐ |
| 5. | | | | ❐ |

# Agenda

| Topic: | Presenter: |
|---|---|
| 1. | |
| 2. | |
| 3. | |
| 4. | |
| 5. | |

| | Action | Person Responsible | Deadline | |
|---|---|---|---|---|
| 1. | | | | ❐ |
| 2. | | | | ❐ |
| 3. | | | | ❐ |
| 4. | | | | ❐ |
| 5. | | | | ❐ |

| Next Meeting Date: | Meeting Time: | Meeting Place: |
|---|---|---|
| | | |

| **Title:** | | |
|---|---|---|
| **Duration:** | **Meeting Date:** | **Meeting Time:** |
| **Location:** | **Called By:** | **Type:** |
| **Facilitator:** | **Note Taker:** | **Time Keeper:** |
| **Attendees:** | | |

## Agenda

| **Topic:** | **Presenter:** |
|---|---|
| 1. | |
| 2. | |
| 3. | |
| 4. | |
| 5. | |

| | Action | Person Responsible | Deadline | |
|---|---|---|---|---|
| 1. | | | | ❒ |
| 2. | | | | ❒ |
| 3. | | | | ❒ |
| 4. | | | | ❒ |
| 5. | | | | ❒ |
| 6. | | | | ❒ |
| 7. | | | | ❒ |
| 8. | | | | ❒ |
| 9. | | | | ❒ |
| 10. | | | | ❒ |

**Notes:**

## Agenda

| Topic: | Presenter: |
|---|---|
| 1. | |
| 2. | |
| 3. | |
| 4. | |
| 5. | |

| | Action | Person Responsible | Deadline | |
|---|---|---|---|---|
| 1. | | | | ❐ |
| 2. | | | | ❐ |
| 3. | | | | ❐ |
| 4. | | | | ❐ |
| 5. | | | | ❐ |

## Agenda

| Topic: | Presenter: |
|---|---|
| 1. | |
| 2. | |
| 3. | |
| 4. | |
| 5. | |

| | Action | Person Responsible | Deadline | |
|---|---|---|---|---|
| 1. | | | | ❐ |
| 2. | | | | ❐ |
| 3. | | | | ❐ |
| 4. | | | | ❐ |
| 5. | | | | ❐ |

| Next Meeting Date: | Meeting Time: | Meeting Place: |
|---|---|---|
| | | |

| **Title:** | | |
|---|---|---|
| **Duration:** | **Meeting Date:** | **Meeting Time:** |
| **Location:** | **Called By:** | **Type:** |
| **Facilitator:** | **Note Taker:** | **Time Keeper:** |
| **Attendees:** | | |

## Agenda

| **Topic:** | **Presenter:** |
|---|---|
| 1. | |
| 2. | |
| 3. | |
| 4. | |
| 5. | |

| | **Action** | **Person Responsible** | **Deadline** | |
|---|---|---|---|---|
| 1. | | | | ❒ |
| 2. | | | | ❒ |
| 3. | | | | ❒ |
| 4. | | | | ❒ |
| 5. | | | | ❒ |
| 6. | | | | ❒ |
| 7. | | | | ❒ |
| 8. | | | | ❒ |
| 9. | | | | ❒ |
| 10. | | | | ❒ |

**Notes:**

| Agenda | |
|---|---|
| Topic: | Presenter: |
| 1. | |
| 2. | |
| 3. | |
| 4. | |
| 5. | |

| | Action | Person Responsible | Deadline | |
|---|---|---|---|---|
| 1. | | | | ❒ |
| 2. | | | | ❒ |
| 3. | | | | ❒ |
| 4. | | | | ❒ |
| 5. | | | | ❒ |

| Agenda | |
|---|---|
| Topic: | Presenter: |
| 1. | |
| 2. | |
| 3. | |
| 4. | |
| 5. | |

| | Action | Person Responsible | Deadline | |
|---|---|---|---|---|
| 1. | | | | ❒ |
| 2. | | | | ❒ |
| 3. | | | | ❒ |
| 4. | | | | ❒ |
| 5. | | | | ❒ |

| Next Meeting Date: | Meeting Time: | Meeting Place: |
|---|---|---|
| | | |

| Title: | | |
|---|---|---|
| Duration: | Meeting Date: | Meeting Time: |
| Location: | Called By: | Type: |
| Facilitator: | Note Taker: | Time Keeper: |
| Attendees: | | |

## Agenda

| Topic: | Presenter: |
|---|---|
| 1. | |
| 2. | |
| 3. | |
| 4. | |
| 5. | |

| | Action | Person Responsible | Deadline | |
|---|---|---|---|---|
| 1. | | | | ❒ |
| 2. | | | | ❒ |
| 3. | | | | ❒ |
| 4. | | | | ❒ |
| 5. | | | | ❒ |
| 6. | | | | ❒ |
| 7. | | | | ❒ |
| 8. | | | | ❒ |
| 9. | | | | ❒ |
| 10. | | | | ❒ |

Notes:

# Agenda

| Topic: | Presenter: |
|---|---|
| 1. | |
| 2. | |
| 3. | |
| 4. | |
| 5. | |

| | Action | Person Responsible | Deadline | |
|---|---|---|---|---|
| 1. | | | | ❒ |
| 2. | | | | ❒ |
| 3. | | | | ❒ |
| 4. | | | | ❒ |
| 5. | | | | ❒ |

# Agenda

| Topic: | Presenter: |
|---|---|
| 1. | |
| 2. | |
| 3. | |
| 4. | |
| 5. | |

| | Action | Person Responsible | Deadline | |
|---|---|---|---|---|
| 1. | | | | ❒ |
| 2. | | | | ❒ |
| 3. | | | | ❒ |
| 4. | | | | ❒ |
| 5. | | | | ❒ |

| Next Meeting Date: | Meeting Time: | Meeting Place: |
|---|---|---|
| | | |

| Title: | | |
|---|---|---|
| Duration: | Meeting Date: | Meeting Time: |
| Location: | Called By: | Type: |
| Facilitator: | Note Taker: | Time Keeper: |
| Attendees: | | |

## Agenda

| Topic: | Presenter: |
|---|---|
| 1. | |
| 2. | |
| 3. | |
| 4. | |
| 5. | |

| | Action | Person Responsible | Deadline | |
|---|---|---|---|---|
| 1. | | | | ❐ |
| 2. | | | | ❐ |
| 3. | | | | ❐ |
| 4. | | | | ❐ |
| 5. | | | | ❐ |
| 6. | | | | ❐ |
| 7. | | | | ❐ |
| 8. | | | | ❐ |
| 9. | | | | ❐ |
| 10. | | | | ❐ |

Notes: ______________________________________________

______________________________________________

______________________________________________

______________________________________________

______________________________________________

______________________________________________

______________________________________________

# Agenda

| Topic: | Presenter: |
|---|---|
| 1. | |
| 2. | |
| 3. | |
| 4. | |
| 5. | |

| | Action | Person Responsible | Deadline | |
|---|---|---|---|---|
| 1. | | | | ❒ |
| 2. | | | | ❒ |
| 3. | | | | ❒ |
| 4. | | | | ❒ |
| 5. | | | | ❒ |

# Agenda

| Topic: | Presenter: |
|---|---|
| 1. | |
| 2. | |
| 3. | |
| 4. | |
| 5. | |

| | Action | Person Responsible | Deadline | |
|---|---|---|---|---|
| 1. | | | | ❒ |
| 2. | | | | ❒ |
| 3. | | | | ❒ |
| 4. | | | | ❒ |
| 5. | | | | ❒ |

| Next Meeting Date: | Meeting Time: | Meeting Place: |
|---|---|---|
| | | |

| **Title:** | | |
|---|---|---|
| **Duration:** | **Meeting Date:** | **Meeting Time:** |
| **Location:** | **Called By:** | **Type:** |
| **Facilitator:** | **Note Taker:** | **Time Keeper:** |
| **Attendees:** | | |

## Agenda

| **Topic:** | **Presenter:** |
|---|---|
| 1. | |
| 2. | |
| 3. | |
| 4. | |
| 5. | |

| | Action | Person Responsible | Deadline | |
|---|---|---|---|---|
| 1. | | | | ❒ |
| 2. | | | | ❒ |
| 3. | | | | ❒ |
| 4. | | | | ❒ |
| 5. | | | | ❒ |
| 6. | | | | ❒ |
| 7. | | | | ❒ |
| 8. | | | | ❒ |
| 9. | | | | ❒ |
| 10. | | | | ❒ |

**Notes:**

# Agenda

| Topic: | Presenter: |
| --- | --- |
| 1. | |
| 2. | |
| 3. | |
| 4. | |
| 5. | |

| | Action | Person Responsible | Deadline | |
| --- | --- | --- | --- | --- |
| 1. | | | | ❒ |
| 2. | | | | ❒ |
| 3. | | | | ❒ |
| 4. | | | | ❒ |
| 5. | | | | ❒ |

# Agenda

| Topic: | Presenter: |
| --- | --- |
| 1. | |
| 2. | |
| 3. | |
| 4. | |
| 5. | |

| | Action | Person Responsible | Deadline | |
| --- | --- | --- | --- | --- |
| 1. | | | | ❒ |
| 2. | | | | ❒ |
| 3. | | | | ❒ |
| 4. | | | | ❒ |
| 5. | | | | ❒ |

| Next Meeting Date: | Meeting Time: | Meeting Place: |
| --- | --- | --- |
| | | |

| Title: | | |
|---|---|---|
| Duration: | Meeting Date: | Meeting Time: |
| Location: | Called By: | Type: |
| Facilitator: | Note Taker: | Time Keeper: |
| Attendees: | | |

## Agenda

| Topic: | Presenter: |
|---|---|
| 1. | |
| 2. | |
| 3. | |
| 4. | |
| 5. | |

| | Action | Person Responsible | Deadline | |
|---|---|---|---|---|
| 1. | | | | ❒ |
| 2. | | | | ❒ |
| 3. | | | | ❒ |
| 4. | | | | ❒ |
| 5. | | | | ❒ |
| 6. | | | | ❒ |
| 7. | | | | ❒ |
| 8. | | | | ❒ |
| 9. | | | | ❒ |
| 10. | | | | ❒ |

**Notes:**

## Agenda

| Topic: | Presenter: |
|---|---|
| 1. | |
| 2. | |
| 3. | |
| 4. | |
| 5. | |

| | Action | Person Responsible | Deadline | |
|---|---|---|---|---|
| 1. | | | | ❐ |
| 2. | | | | ❐ |
| 3. | | | | ❐ |
| 4. | | | | ❐ |
| 5. | | | | ❐ |

## Agenda

| Topic: | Presenter: |
|---|---|
| 1. | |
| 2. | |
| 3. | |
| 4. | |
| 5. | |

| | Action | Person Responsible | Deadline | |
|---|---|---|---|---|
| 1. | | | | ❐ |
| 2. | | | | ❐ |
| 3. | | | | ❐ |
| 4. | | | | ❐ |
| 5. | | | | ❐ |

| Next Meeting Date: | Meeting Time: | Meeting Place: |
|---|---|---|
| | | |

| Title: | | |
|---|---|---|
| Duration: | Meeting Date: | Meeting Time: |
| Location: | Called By: | Type: |
| Facilitator: | Note Taker: | Time Keeper: |
| Attendees: | | |

## Agenda

| Topic: | Presenter: |
|---|---|
| 1. | |
| 2. | |
| 3. | |
| 4. | |
| 5. | |

| | Action | Person Responsible | Deadline | |
|---|---|---|---|---|
| 1. | | | | ❒ |
| 2. | | | | ❒ |
| 3. | | | | ❒ |
| 4. | | | | ❒ |
| 5. | | | | ❒ |
| 6. | | | | ❒ |
| 7. | | | | ❒ |
| 8. | | | | ❒ |
| 9. | | | | ❒ |
| 10. | | | | ❒ |

**Notes:**

| Agenda | |
|---|---|
| **Topic:** | **Presenter:** |
| 1. | |
| 2. | |
| 3. | |
| 4. | |
| 5. | |

| | **Action** | **Person Responsible** | **Deadline** | |
|---|---|---|---|---|
| 1. | | | | ❐ |
| 2. | | | | ❐ |
| 3. | | | | ❐ |
| 4. | | | | ❐ |
| 5. | | | | ❐ |

| Agenda | |
|---|---|
| **Topic:** | **Presenter:** |
| 1. | |
| 2. | |
| 3. | |
| 4. | |
| 5. | |

| | **Action** | **Person Responsible** | **Deadline** | |
|---|---|---|---|---|
| 1. | | | | ❐ |
| 2. | | | | ❐ |
| 3. | | | | ❐ |
| 4. | | | | ❐ |
| 5. | | | | ❐ |

| **Next Meeting Date:** | **Meeting Time:** | **Meeting Place:** |
|---|---|---|
| | | |

| Title: | | |
|---|---|---|
| Duration: | Meeting Date: | Meeting Time: |
| Location: | Called By: | Type: |
| Facilitator: | Note Taker: | Time Keeper: |
| Attendees: | | |

## Agenda

| Topic: | Presenter: |
|---|---|
| 1. | |
| 2. | |
| 3. | |
| 4. | |
| 5. | |

| | Action | Person Responsible | Deadline | |
|---|---|---|---|---|
| 1. | | | | ❐ |
| 2. | | | | ❐ |
| 3. | | | | ❐ |
| 4. | | | | ❐ |
| 5. | | | | ❐ |
| 6. | | | | ❐ |
| 7. | | | | ❐ |
| 8. | | | | ❐ |
| 9. | | | | ❐ |
| 10. | | | | ❐ |

Notes: ______________________________

______________________________

______________________________

______________________________

______________________________

______________________________

______________________________

# Agenda

| Topic: | Presenter: |
|---|---|
| 1. | |
| 2. | |
| 3. | |
| 4. | |
| 5. | |

| | Action | Person Responsible | Deadline | |
|---|---|---|---|---|
| 1. | | | | ❐ |
| 2. | | | | ❐ |
| 3. | | | | ❐ |
| 4. | | | | ❐ |
| 5. | | | | ❐ |

# Agenda

| Topic: | Presenter: |
|---|---|
| 1. | |
| 2. | |
| 3. | |
| 4. | |
| 5. | |

| | Action | Person Responsible | Deadline | |
|---|---|---|---|---|
| 1. | | | | ❐ |
| 2. | | | | ❐ |
| 3. | | | | ❐ |
| 4. | | | | ❐ |
| 5. | | | | ❐ |

| Next Meeting Date: | Meeting Time: | Meeting Place: |
|---|---|---|
| | | |

| **Title:** | | |
|---|---|---|
| **Duration:** | **Meeting Date:** | **Meeting Time:** |
| **Location:** | **Called By:** | **Type:** |
| **Facilitator:** | **Note Taker:** | **Time Keeper:** |
| **Attendees:** | | |

## Agenda

| **Topic:** | **Presenter:** |
|---|---|
| 1. | |
| 2. | |
| 3. | |
| 4. | |
| 5. | |

| | **Action** | **Person Responsible** | **Deadline** | |
|---|---|---|---|---|
| 1. | | | | ❒ |
| 2. | | | | ❒ |
| 3. | | | | ❒ |
| 4. | | | | ❒ |
| 5. | | | | ❒ |
| 6. | | | | ❒ |
| 7. | | | | ❒ |
| 8. | | | | ❒ |
| 9. | | | | ❒ |
| 10. | | | | ❒ |

**Notes:**

| Agenda | |
|---|---|
| **Topic:** | **Presenter:** |
| 1. | |
| 2. | |
| 3. | |
| 4. | |
| 5. | |

| | Action | Person Responsible | Deadline | |
|---|---|---|---|---|
| 1. | | | | ❐ |
| 2. | | | | ❐ |
| 3. | | | | ❐ |
| 4. | | | | ❐ |
| 5. | | | | ❐ |

| Agenda | |
|---|---|
| **Topic:** | **Presenter:** |
| 1. | |
| 2. | |
| 3. | |
| 4. | |
| 5. | |

| | Action | Person Responsible | Deadline | |
|---|---|---|---|---|
| 1. | | | | ❐ |
| 2. | | | | ❐ |
| 3. | | | | ❐ |
| 4. | | | | ❐ |
| 5. | | | | ❐ |

| Next Meeting Date: | Meeting Time: | Meeting Place: |
|---|---|---|
| | | |

| **Title:** | | |
|---|---|---|
| **Duration:** | **Meeting Date:** | **Meeting Time:** |
| **Location:** | **Called By:** | **Type:** |
| **Facilitator:** | **Note Taker:** | **Time Keeper:** |
| **Attendees:** | | |

## Agenda

| **Topic:** | **Presenter:** |
|---|---|
| 1. | |
| 2. | |
| 3. | |
| 4. | |
| 5. | |

| | **Action** | **Person Responsible** | **Deadline** | |
|---|---|---|---|---|
| 1. | | | | ❒ |
| 2. | | | | ❒ |
| 3. | | | | ❒ |
| 4. | | | | ❒ |
| 5. | | | | ❒ |
| 6. | | | | ❒ |
| 7. | | | | ❒ |
| 8. | | | | ❒ |
| 9. | | | | ❒ |
| 10. | | | | ❒ |

**Notes:**

# Agenda

| Topic: | Presenter: |
|---|---|
| 1. | |
| 2. | |
| 3. | |
| 4. | |
| 5. | |

| | Action | Person Responsible | Deadline | |
|---|---|---|---|---|
| 1. | | | | ❒ |
| 2. | | | | ❒ |
| 3. | | | | ❒ |
| 4. | | | | ❒ |
| 5. | | | | ❒ |

# Agenda

| Topic: | Presenter: |
|---|---|
| 1. | |
| 2. | |
| 3. | |
| 4. | |
| 5. | |

| | Action | Person Responsible | Deadline | |
|---|---|---|---|---|
| 1. | | | | ❒ |
| 2. | | | | ❒ |
| 3. | | | | ❒ |
| 4. | | | | ❒ |
| 5. | | | | ❒ |

| Next Meeting Date: | Meeting Time: | Meeting Place: |
|---|---|---|
| | | |

| Title: | | |
|---|---|---|
| Duration: | Meeting Date: | Meeting Time: |
| Location: | Called By: | Type: |
| Facilitator: | Note Taker: | Time Keeper: |
| Attendees: | | |

## Agenda

| Topic: | Presenter: |
|---|---|
| 1. | |
| 2. | |
| 3. | |
| 4. | |
| 5. | |

| | Action | Person Responsible | Deadline | |
|---|---|---|---|---|
| 1. | | | | ❒ |
| 2. | | | | ❒ |
| 3. | | | | ❒ |
| 4. | | | | ❒ |
| 5. | | | | ❒ |
| 6. | | | | ❒ |
| 7. | | | | ❒ |
| 8. | | | | ❒ |
| 9. | | | | ❒ |
| 10. | | | | ❒ |

Notes:

# Agenda

| Topic: | Presenter: |
|---|---|
| 1. | |
| 2. | |
| 3. | |
| 4. | |
| 5. | |

| | Action | Person Responsible | Deadline | |
|---|---|---|---|---|
| 1. | | | | ❒ |
| 2. | | | | ❒ |
| 3. | | | | ❒ |
| 4. | | | | ❒ |
| 5. | | | | ❒ |

# Agenda

| Topic: | Presenter: |
|---|---|
| 1. | |
| 2. | |
| 3. | |
| 4. | |
| 5. | |

| | Action | Person Responsible | Deadline | |
|---|---|---|---|---|
| 1. | | | | ❒ |
| 2. | | | | ❒ |
| 3. | | | | ❒ |
| 4. | | | | ❒ |
| 5. | | | | ❒ |

| Next Meeting Date: | Meeting Time: | Meeting Place: |
|---|---|---|
| | | |

| Title: | | |
|---|---|---|
| Duration: | Meeting Date: | Meeting Time: |
| Location: | Called By: | Type: |
| Facilitator: | Note Taker: | Time Keeper: |
| Attendees: | | |

## Agenda

| Topic: | Presenter: |
|---|---|
| 1. | |
| 2. | |
| 3. | |
| 4. | |
| 5. | |

| | Action | Person Responsible | Deadline | |
|---|---|---|---|---|
| 1. | | | | ❒ |
| 2. | | | | ❒ |
| 3. | | | | ❒ |
| 4. | | | | ❒ |
| 5. | | | | ❒ |
| 6. | | | | ❒ |
| 7. | | | | ❒ |
| 8. | | | | ❒ |
| 9. | | | | ❒ |
| 10. | | | | ❒ |

**Notes:**

# Agenda

| Topic: | Presenter: |
|---|---|
| 1. | |
| 2. | |
| 3. | |
| 4. | |
| 5. | |

| | Action | Person Responsible | Deadline | |
|---|---|---|---|---|
| 1. | | | | ❐ |
| 2. | | | | ❐ |
| 3. | | | | ❐ |
| 4. | | | | ❐ |
| 5. | | | | ❐ |

# Agenda

| Topic: | Presenter: |
|---|---|
| 1. | |
| 2. | |
| 3. | |
| 4. | |
| 5. | |

| | Action | Person Responsible | Deadline | |
|---|---|---|---|---|
| 1. | | | | ❐ |
| 2. | | | | ❐ |
| 3. | | | | ❐ |
| 4. | | | | ❐ |
| 5. | | | | ❐ |

| Next Meeting Date: | Meeting Time: | Meeting Place: |
|---|---|---|
| | | |

| Title: | | |
|---|---|---|
| Duration: | Meeting Date: | Meeting Time: |
| Location: | Called By: | Type: |
| Facilitator: | Note Taker: | Time Keeper: |
| Attendees: | | |

## Agenda

| Topic: | Presenter: |
|---|---|
| 1. | |
| 2. | |
| 3. | |
| 4. | |
| 5. | |

| | Action | Person Responsible | Deadline | |
|---|---|---|---|---|
| 1. | | | | ❐ |
| 2. | | | | ❐ |
| 3. | | | | ❐ |
| 4. | | | | ❐ |
| 5. | | | | ❐ |
| 6. | | | | ❐ |
| 7. | | | | ❐ |
| 8. | | | | ❐ |
| 9. | | | | ❐ |
| 10. | | | | ❐ |

**Notes:**

| Agenda | |
|---|---|
| Topic: | Presenter: |
| 1. | |
| 2. | |
| 3. | |
| 4. | |
| 5. | |

| | Action | Person Responsible | Deadline | |
|---|---|---|---|---|
| 1. | | | | ❒ |
| 2. | | | | ❒ |
| 3. | | | | ❒ |
| 4. | | | | ❒ |
| 5. | | | | ❒ |

| Agenda | |
|---|---|
| Topic: | Presenter: |
| 1. | |
| 2. | |
| 3. | |
| 4. | |
| 5. | |

| | Action | Person Responsible | Deadline | |
|---|---|---|---|---|
| 1. | | | | ❒ |
| 2. | | | | ❒ |
| 3. | | | | ❒ |
| 4. | | | | ❒ |
| 5. | | | | ❒ |

| Next Meeting Date: | Meeting Time: | Meeting Place: |
|---|---|---|
| | | |

| **Title:** | | |
|---|---|---|
| **Duration:** | **Meeting Date:** | **Meeting Time:** |
| **Location:** | **Called By:** | **Type:** |
| **Facilitator:** | **Note Taker:** | **Time Keeper:** |
| **Attendees:** | | |

## Agenda

| **Topic:** | **Presenter:** |
|---|---|
| 1. | |
| 2. | |
| 3. | |
| 4. | |
| 5. | |

| | **Action** | **Person Responsible** | **Deadline** | |
|---|---|---|---|---|
| 1. | | | | ❐ |
| 2. | | | | ❐ |
| 3. | | | | ❐ |
| 4. | | | | ❐ |
| 5. | | | | ❐ |
| 6. | | | | ❐ |
| 7. | | | | ❐ |
| 8. | | | | ❐ |
| 9. | | | | ❐ |
| 10. | | | | ❐ |

**Notes:**

## Agenda

| Topic: | Presenter: |
|---|---|
| 1. | |
| 2. | |
| 3. | |
| 4. | |
| 5. | |

| | Action | Person Responsible | Deadline | |
|---|---|---|---|---|
| 1. | | | | ❐ |
| 2. | | | | ❐ |
| 3. | | | | ❐ |
| 4. | | | | ❐ |
| 5. | | | | ❐ |

## Agenda

| Topic: | Presenter: |
|---|---|
| 1. | |
| 2. | |
| 3. | |
| 4. | |
| 5. | |

| | Action | Person Responsible | Deadline | |
|---|---|---|---|---|
| 1. | | | | ❐ |
| 2. | | | | ❐ |
| 3. | | | | ❐ |
| 4. | | | | ❐ |
| 5. | | | | ❐ |

| Next Meeting Date: | Meeting Time: | Meeting Place: |
|---|---|---|
| | | |

| **Title:** | | |
|---|---|---|
| **Duration:** | **Meeting Date:** | **Meeting Time:** |
| **Location:** | **Called By:** | **Type:** |
| **Facilitator:** | **Note Taker:** | **Time Keeper:** |
| **Attendees:** | | |

## Agenda

| **Topic:** | **Presenter:** |
|---|---|
| 1. | |
| 2. | |
| 3. | |
| 4. | |
| 5. | |

| | **Action** | **Person Responsible** | **Deadline** | |
|---|---|---|---|---|
| 1. | | | | ❐ |
| 2. | | | | ❐ |
| 3. | | | | ❐ |
| 4. | | | | ❐ |
| 5. | | | | ❐ |
| 6. | | | | ❐ |
| 7. | | | | ❐ |
| 8. | | | | ❐ |
| 9. | | | | ❐ |
| 10. | | | | ❐ |

**Notes:**

| Agenda | |
|---|---|
| **Topic:** | **Presenter:** |
| 1. | |
| 2. | |
| 3. | |
| 4. | |
| 5. | |

| | Action | Person Responsible | Deadline | |
|---|---|---|---|---|
| 1. | | | | ❐ |
| 2. | | | | ❐ |
| 3. | | | | ❐ |
| 4. | | | | ❐ |
| 5. | | | | ❐ |

| Agenda | |
|---|---|
| **Topic:** | **Presenter:** |
| 1. | |
| 2. | |
| 3. | |
| 4. | |
| 5. | |

| | Action | Person Responsible | Deadline | |
|---|---|---|---|---|
| 1. | | | | ❐ |
| 2. | | | | ❐ |
| 3. | | | | ❐ |
| 4. | | | | ❐ |
| 5. | | | | ❐ |

| Next Meeting Date: | Meeting Time: | Meeting Place: |
|---|---|---|
| | | |

| Title: | | |
|---|---|---|
| Duration: | Meeting Date: | Meeting Time: |
| Location: | Called By: | Type: |
| Facilitator: | Note Taker: | Time Keeper: |
| Attendees: | | |

## Agenda

| Topic: | Presenter: |
|---|---|
| 1. | |
| 2. | |
| 3. | |
| 4. | |
| 5. | |

| | Action | Person Responsible | Deadline | |
|---|---|---|---|---|
| 1. | | | | ❐ |
| 2. | | | | ❐ |
| 3. | | | | ❐ |
| 4. | | | | ❐ |
| 5. | | | | ❐ |
| 6. | | | | ❐ |
| 7. | | | | ❐ |
| 8. | | | | ❐ |
| 9. | | | | ❐ |
| 10. | | | | ❐ |

**Notes:**

| Agenda | |
|---|---|
| Topic: | Presenter: |
| 1. | |
| 2. | |
| 3. | |
| 4. | |
| 5. | |

| | Action | Person Responsible | Deadline | |
|---|---|---|---|---|
| 1. | | | | ❐ |
| 2. | | | | ❐ |
| 3. | | | | ❐ |
| 4. | | | | ❐ |
| 5. | | | | ❐ |

| Agenda | |
|---|---|
| Topic: | Presenter: |
| 1. | |
| 2. | |
| 3. | |
| 4. | |
| 5. | |

| | Action | Person Responsible | Deadline | |
|---|---|---|---|---|
| 1. | | | | ❐ |
| 2. | | | | ❐ |
| 3. | | | | ❐ |
| 4. | | | | ❐ |
| 5. | | | | ❐ |

| Next Meeting Date: | Meeting Time: | Meeting Place: |
|---|---|---|
| | | |

| **Title:** | | |
|---|---|---|
| **Duration:** | **Meeting Date:** | **Meeting Time:** |
| **Location:** | **Called By:** | **Type:** |
| **Facilitator:** | **Note Taker:** | **Time Keeper:** |
| **Attendees:** | | |

## Agenda

| **Topic:** | **Presenter:** |
|---|---|
| 1. | |
| 2. | |
| 3. | |
| 4. | |
| 5. | |

| | Action | Person Responsible | Deadline | |
|---|---|---|---|---|
| 1. | | | | ❐ |
| 2. | | | | ❐ |
| 3. | | | | ❐ |
| 4. | | | | ❐ |
| 5. | | | | ❐ |
| 6. | | | | ❐ |
| 7. | | | | ❐ |
| 8. | | | | ❐ |
| 9. | | | | ❐ |
| 10. | | | | ❐ |

**Notes:**

## Agenda

| Topic: | Presenter: |
|---|---|
| 1. | |
| 2. | |
| 3. | |
| 4. | |
| 5. | |

| | Action | Person Responsible | Deadline | |
|---|---|---|---|---|
| 1. | | | | ❐ |
| 2. | | | | ❐ |
| 3. | | | | ❐ |
| 4. | | | | ❐ |
| 5. | | | | ❐ |

## Agenda

| Topic: | Presenter: |
|---|---|
| 1. | |
| 2. | |
| 3. | |
| 4. | |
| 5. | |

| | Action | Person Responsible | Deadline | |
|---|---|---|---|---|
| 1. | | | | ❐ |
| 2. | | | | ❐ |
| 3. | | | | ❐ |
| 4. | | | | ❐ |
| 5. | | | | ❐ |

| Next Meeting Date: | Meeting Time: | Meeting Place: |
|---|---|---|
| | | |

| Title: | | |
|---|---|---|
| Duration: | Meeting Date: | Meeting Time: |
| Location: | Called By: | Type: |
| Facilitator: | Note Taker: | Time Keeper: |
| Attendees: | | |

## Agenda

| Topic: | Presenter: |
|---|---|
| 1. | |
| 2. | |
| 3. | |
| 4. | |
| 5. | |

| | Action | Person Responsible | Deadline | |
|---|---|---|---|---|
| 1. | | | | ❐ |
| 2. | | | | ❐ |
| 3. | | | | ❐ |
| 4. | | | | ❐ |
| 5. | | | | ❐ |
| 6. | | | | ❐ |
| 7. | | | | ❐ |
| 8. | | | | ❐ |
| 9. | | | | ❐ |
| 10. | | | | ❐ |

**Notes:**

# Agenda

| Topic: | Presenter: |
|---|---|
| 1. | |
| 2. | |
| 3. | |
| 4. | |
| 5. | |

| | Action | Person Responsible | Deadline | |
|---|---|---|---|---|
| 1. | | | | ❒ |
| 2. | | | | ❒ |
| 3. | | | | ❒ |
| 4. | | | | ❒ |
| 5. | | | | ❒ |

# Agenda

| Topic: | Presenter: |
|---|---|
| 1. | |
| 2. | |
| 3. | |
| 4. | |
| 5. | |

| | Action | Person Responsible | Deadline | |
|---|---|---|---|---|
| 1. | | | | ❒ |
| 2. | | | | ❒ |
| 3. | | | | ❒ |
| 4. | | | | ❒ |
| 5. | | | | ❒ |

| Next Meeting Date: | Meeting Time: | Meeting Place: |
|---|---|---|
| | | |

| Title: | | |
|---|---|---|
| Duration: | Meeting Date: | Meeting Time: |
| Location: | Called By: | Type: |
| Facilitator: | Note Taker: | Time Keeper: |
| Attendees: | | |

## Agenda

| Topic: | Presenter: |
|---|---|
| 1. | |
| 2. | |
| 3. | |
| 4. | |
| 5. | |

| | Action | Person Responsible | Deadline | |
|---|---|---|---|---|
| 1. | | | | ❒ |
| 2. | | | | ❒ |
| 3. | | | | ❒ |
| 4. | | | | ❒ |
| 5. | | | | ❒ |
| 6. | | | | ❒ |
| 7. | | | | ❒ |
| 8. | | | | ❒ |
| 9. | | | | ❒ |
| 10. | | | | ❒ |

**Notes:**

| Agenda | |
|---|---|
| **Topic:** | **Presenter:** |
| 1. | |
| 2. | |
| 3. | |
| 4. | |
| 5. | |

| | Action | Person Responsible | Deadline | |
|---|---|---|---|---|
| 1. | | | | ❐ |
| 2. | | | | ❐ |
| 3. | | | | ❐ |
| 4. | | | | ❐ |
| 5. | | | | ❐ |

| Agenda | |
|---|---|
| **Topic:** | **Presenter:** |
| 1. | |
| 2. | |
| 3. | |
| 4. | |
| 5. | |

| | Action | Person Responsible | Deadline | |
|---|---|---|---|---|
| 1. | | | | ❐ |
| 2. | | | | ❐ |
| 3. | | | | ❐ |
| 4. | | | | ❐ |
| 5. | | | | ❐ |

| Next Meeting Date: | Meeting Time: | Meeting Place: |
|---|---|---|
| | | |

| **Title:** | | |
|---|---|---|
| **Duration:** | **Meeting Date:** | **Meeting Time:** |
| **Location:** | **Called By:** | **Type:** |
| **Facilitator:** | **Note Taker:** | **Time Keeper:** |
| **Attendees:** | | |

## Agenda

| **Topic:** | **Presenter:** |
|---|---|
| 1. | |
| 2. | |
| 3. | |
| 4. | |
| 5. | |

| | Action | Person Responsible | Deadline | |
|---|---|---|---|---|
| 1. | | | | ❐ |
| 2. | | | | ❐ |
| 3. | | | | ❐ |
| 4. | | | | ❐ |
| 5. | | | | ❐ |
| 6. | | | | ❐ |
| 7. | | | | ❐ |
| 8. | | | | ❐ |
| 9. | | | | ❐ |
| 10. | | | | ❐ |

**Notes:**

# Agenda

| Topic: | Presenter: |
|---|---|
| 1. | |
| 2. | |
| 3. | |
| 4. | |
| 5. | |

| | Action | Person Responsible | Deadline | |
|---|---|---|---|---|
| 1. | | | | ❒ |
| 2. | | | | ❒ |
| 3. | | | | ❒ |
| 4. | | | | ❒ |
| 5. | | | | ❒ |

# Agenda

| Topic: | Presenter: |
|---|---|
| 1. | |
| 2. | |
| 3. | |
| 4. | |
| 5. | |

| | Action | Person Responsible | Deadline | |
|---|---|---|---|---|
| 1. | | | | ❒ |
| 2. | | | | ❒ |
| 3. | | | | ❒ |
| 4. | | | | ❒ |
| 5. | | | | ❒ |

| Next Meeting Date: | Meeting Time: | Meeting Place: |
|---|---|---|
| | | |

| **Title:** | | |
|---|---|---|
| **Duration:** | **Meeting Date:** | **Meeting Time:** |
| **Location:** | **Called By:** | **Type:** |
| **Facilitator:** | **Note Taker:** | **Time Keeper:** |
| **Attendees:** | | |

## Agenda

| **Topic:** | **Presenter:** |
|---|---|
| 1. | |
| 2. | |
| 3. | |
| 4. | |
| 5. | |

| | **Action** | **Person Responsible** | **Deadline** | |
|---|---|---|---|---|
| 1. | | | | ❐ |
| 2. | | | | ❐ |
| 3. | | | | ❐ |
| 4. | | | | ❐ |
| 5. | | | | ❐ |
| 6. | | | | ❐ |
| 7. | | | | ❐ |
| 8. | | | | ❐ |
| 9. | | | | ❐ |
| 10. | | | | ❐ |

**Notes:**

## Agenda

| Topic: | Presenter: |
|---|---|
| 1. | |
| 2. | |
| 3. | |
| 4. | |
| 5. | |

| | Action | Person Responsible | Deadline | |
|---|---|---|---|---|
| 1. | | | | ❒ |
| 2. | | | | ❒ |
| 3. | | | | ❒ |
| 4. | | | | ❒ |
| 5. | | | | ❒ |

## Agenda

| Topic: | Presenter: |
|---|---|
| 1. | |
| 2. | |
| 3. | |
| 4. | |
| 5. | |

| | Action | Person Responsible | Deadline | |
|---|---|---|---|---|
| 1. | | | | ❒ |
| 2. | | | | ❒ |
| 3. | | | | ❒ |
| 4. | | | | ❒ |
| 5. | | | | ❒ |

| Next Meeting Date: | Meeting Time: | Meeting Place: |
|---|---|---|
| | | |

| **Title:** | | |
|---|---|---|
| **Duration:** | **Meeting Date:** | **Meeting Time:** |
| **Location:** | **Called By:** | **Type:** |
| **Facilitator:** | **Note Taker:** | **Time Keeper:** |
| **Attendees:** | | |

## Agenda

| **Topic:** | **Presenter:** |
|---|---|
| 1. | |
| 2. | |
| 3. | |
| 4. | |
| 5. | |

| | **Action** | **Person Responsible** | **Deadline** | |
|---|---|---|---|---|
| 1. | | | | ❒ |
| 2. | | | | ❒ |
| 3. | | | | ❒ |
| 4. | | | | ❒ |
| 5. | | | | ❒ |
| 6. | | | | ❒ |
| 7. | | | | ❒ |
| 8. | | | | ❒ |
| 9. | | | | ❒ |
| 10. | | | | ❒ |

**Notes:**

# Agenda

| Topic: | Presenter: |
|---|---|
| 1. | |
| 2. | |
| 3. | |
| 4. | |
| 5. | |

| | Action | Person Responsible | Deadline | |
|---|---|---|---|---|
| 1. | | | | ❐ |
| 2. | | | | ❐ |
| 3. | | | | ❐ |
| 4. | | | | ❐ |
| 5. | | | | ❐ |

# Agenda

| Topic: | Presenter: |
|---|---|
| 1. | |
| 2. | |
| 3. | |
| 4. | |
| 5. | |

| | Action | Person Responsible | Deadline | |
|---|---|---|---|---|
| 1. | | | | ❐ |
| 2. | | | | ❐ |
| 3. | | | | ❐ |
| 4. | | | | ❐ |
| 5. | | | | ❐ |

| Next Meeting Date: | Meeting Time: | Meeting Place: |
|---|---|---|
| | | |

| **Title:** | | |
|---|---|---|
| **Duration:** | **Meeting Date:** | **Meeting Time:** |
| **Location:** | **Called By:** | **Type:** |
| **Facilitator:** | **Note Taker:** | **Time Keeper:** |
| **Attendees:** | | |

## Agenda

| **Topic:** | **Presenter:** |
|---|---|
| 1. | |
| 2. | |
| 3. | |
| 4. | |
| 5. | |

| | Action | Person Responsible | Deadline | |
|---|---|---|---|---|
| 1. | | | | ❒ |
| 2. | | | | ❒ |
| 3. | | | | ❒ |
| 4. | | | | ❒ |
| 5. | | | | ❒ |
| 6. | | | | ❒ |
| 7. | | | | ❒ |
| 8. | | | | ❒ |
| 9. | | | | ❒ |
| 10. | | | | ❒ |

**Notes:**

# Agenda

| Topic: | Presenter: |
|---|---|
| 1. | |
| 2. | |
| 3. | |
| 4. | |
| 5. | |

| | Action | Person Responsible | Deadline | |
|---|---|---|---|---|
| 1. | | | | ❐ |
| 2. | | | | ❐ |
| 3. | | | | ❐ |
| 4. | | | | ❐ |
| 5. | | | | ❐ |

# Agenda

| Topic: | Presenter: |
|---|---|
| 1. | |
| 2. | |
| 3. | |
| 4. | |
| 5. | |

| | Action | Person Responsible | Deadline | |
|---|---|---|---|---|
| 1. | | | | ❐ |
| 2. | | | | ❐ |
| 3. | | | | ❐ |
| 4. | | | | ❐ |
| 5. | | | | ❐ |

| Next Meeting Date: | Meeting Time: | Meeting Place: |
|---|---|---|
| | | |

| Title: | | |
|---|---|---|
| Duration: | Meeting Date: | Meeting Time: |
| Location: | Called By: | Type: |
| Facilitator: | Note Taker: | Time Keeper: |
| Attendees: | | |

| Agenda | |
|---|---|
| Topic: | Presenter: |
| 1. | |
| 2. | |
| 3. | |
| 4. | |
| 5. | |

| | Action | Person Responsible | Deadline | |
|---|---|---|---|---|
| 1. | | | | ❒ |
| 2. | | | | ❒ |
| 3. | | | | ❒ |
| 4. | | | | ❒ |
| 5. | | | | ❒ |
| 6. | | | | ❒ |
| 7. | | | | ❒ |
| 8. | | | | ❒ |
| 9. | | | | ❒ |
| 10. | | | | ❒ |

Notes:

| Agenda | |
|---|---|
| Topic: | Presenter: |
| 1. | |
| 2. | |
| 3. | |
| 4. | |
| 5. | |

| | Action | Person Responsible | Deadline | |
|---|---|---|---|---|
| 1. | | | | ❒ |
| 2. | | | | ❒ |
| 3. | | | | ❒ |
| 4. | | | | ❒ |
| 5. | | | | ❒ |

| Agenda | |
|---|---|
| Topic: | Presenter: |
| 1. | |
| 2. | |
| 3. | |
| 4. | |
| 5. | |

| | Action | Person Responsible | Deadline | |
|---|---|---|---|---|
| 1. | | | | ❒ |
| 2. | | | | ❒ |
| 3. | | | | ❒ |
| 4. | | | | ❒ |
| 5. | | | | ❒ |

| Next Meeting Date: | Meeting Time: | Meeting Place: |
|---|---|---|
| | | |

| Title: | | |
|---|---|---|
| Duration: | Meeting Date: | Meeting Time: |
| Location: | Called By: | Type: |
| Facilitator: | Note Taker: | Time Keeper: |
| Attendees: | | |

## Agenda

| Topic: | Presenter: |
|---|---|
| 1. | |
| 2. | |
| 3. | |
| 4. | |
| 5. | |

| | Action | Person Responsible | Deadline | |
|---|---|---|---|---|
| 1. | | | | ❒ |
| 2. | | | | ❒ |
| 3. | | | | ❒ |
| 4. | | | | ❒ |
| 5. | | | | ❒ |
| 6. | | | | ❒ |
| 7. | | | | ❒ |
| 8. | | | | ❒ |
| 9. | | | | ❒ |
| 10. | | | | ❒ |

**Notes:**

## Agenda

| Topic: | Presenter: |
|---|---|
| 1. | |
| 2. | |
| 3. | |
| 4. | |
| 5. | |

| | Action | Person Responsible | Deadline | |
|---|---|---|---|---|
| 1. | | | | ❒ |
| 2. | | | | ❒ |
| 3. | | | | ❒ |
| 4. | | | | ❒ |
| 5. | | | | ❒ |

## Agenda

| Topic: | Presenter: |
|---|---|
| 1. | |
| 2. | |
| 3. | |
| 4. | |
| 5. | |

| | Action | Person Responsible | Deadline | |
|---|---|---|---|---|
| 1. | | | | ❒ |
| 2. | | | | ❒ |
| 3. | | | | ❒ |
| 4. | | | | ❒ |
| 5. | | | | ❒ |

| Next Meeting Date: | Meeting Time: | Meeting Place: |
|---|---|---|
| | | |

| Title: | | |
|---|---|---|
| Duration: | Meeting Date: | Meeting Time: |
| Location: | Called By: | Type: |
| Facilitator: | Note Taker: | Time Keeper: |
| Attendees: | | |

## Agenda

| Topic: | Presenter: |
|---|---|
| 1. | |
| 2. | |
| 3. | |
| 4. | |
| 5. | |

| | Action | Person Responsible | Deadline | |
|---|---|---|---|---|
| 1. | | | | ❐ |
| 2. | | | | ❐ |
| 3. | | | | ❐ |
| 4. | | | | ❐ |
| 5. | | | | ❐ |
| 6. | | | | ❐ |
| 7. | | | | ❐ |
| 8. | | | | ❐ |
| 9. | | | | ❐ |
| 10. | | | | ❐ |

**Notes:**

# Agenda

| Topic: | Presenter: |
|---|---|
| 1. | |
| 2. | |
| 3. | |
| 4. | |
| 5. | |

| | Action | Person Responsible | Deadline | |
|---|---|---|---|---|
| 1. | | | | ❐ |
| 2. | | | | ❐ |
| 3. | | | | ❐ |
| 4. | | | | ❐ |
| 5. | | | | ❐ |

# Agenda

| Topic: | Presenter: |
|---|---|
| 1. | |
| 2. | |
| 3. | |
| 4. | |
| 5. | |

| | Action | Person Responsible | Deadline | |
|---|---|---|---|---|
| 1. | | | | ❐ |
| 2. | | | | ❐ |
| 3. | | | | ❐ |
| 4. | | | | ❐ |
| 5. | | | | ❐ |

| Next Meeting Date: | Meeting Time: | Meeting Place: |
|---|---|---|
| | | |

| Title: | | |
|---|---|---|
| Duration: | Meeting Date: | Meeting Time: |
| Location: | Called By: | Type: |
| Facilitator: | Note Taker: | Time Keeper: |
| Attendees: | | |

## Agenda

| Topic: | Presenter: |
|---|---|
| 1. | |
| 2. | |
| 3. | |
| 4. | |
| 5. | |

| | Action | Person Responsible | Deadline | |
|---|---|---|---|---|
| 1. | | | | ❐ |
| 2. | | | | ❐ |
| 3. | | | | ❐ |
| 4. | | | | ❐ |
| 5. | | | | ❐ |
| 6. | | | | ❐ |
| 7. | | | | ❐ |
| 8. | | | | ❐ |
| 9. | | | | ❐ |
| 10. | | | | ❐ |

Notes:

## Agenda

| Topic: | Presenter: |
|---|---|
| 1. | |
| 2. | |
| 3. | |
| 4. | |
| 5. | |

| | Action | Person Responsible | Deadline | |
|---|---|---|---|---|
| 1. | | | | ❐ |
| 2. | | | | ❐ |
| 3. | | | | ❐ |
| 4. | | | | ❐ |
| 5. | | | | ❐ |

## Agenda

| Topic: | Presenter: |
|---|---|
| 1. | |
| 2. | |
| 3. | |
| 4. | |
| 5. | |

| | Action | Person Responsible | Deadline | |
|---|---|---|---|---|
| 1. | | | | ❐ |
| 2. | | | | ❐ |
| 3. | | | | ❐ |
| 4. | | | | ❐ |
| 5. | | | | ❐ |

| Next Meeting Date: | Meeting Time: | Meeting Place: |
|---|---|---|
| | | |

| Title: | | |
|---|---|---|
| Duration: | Meeting Date: | Meeting Time: |
| Location: | Called By: | Type: |
| Facilitator: | Note Taker: | Time Keeper: |
| Attendees: | | |

## Agenda

| Topic: | Presenter: |
|---|---|
| 1. | |
| 2. | |
| 3. | |
| 4. | |
| 5. | |

| | Action | Person Responsible | Deadline | |
|---|---|---|---|---|
| 1. | | | | ❒ |
| 2. | | | | ❒ |
| 3. | | | | ❒ |
| 4. | | | | ❒ |
| 5. | | | | ❒ |
| 6. | | | | ❒ |
| 7. | | | | ❒ |
| 8. | | | | ❒ |
| 9. | | | | ❒ |
| 10. | | | | ❒ |

**Notes:**

## Agenda

| Topic: | Presenter: |
|---|---|
| 1. | |
| 2. | |
| 3. | |
| 4. | |
| 5. | |

| | Action | Person Responsible | Deadline | |
|---|---|---|---|---|
| 1. | | | | ❒ |
| 2. | | | | ❒ |
| 3. | | | | ❒ |
| 4. | | | | ❒ |
| 5. | | | | ❒ |

## Agenda

| Topic: | Presenter: |
|---|---|
| 1. | |
| 2. | |
| 3. | |
| 4. | |
| 5. | |

| | Action | Person Responsible | Deadline | |
|---|---|---|---|---|
| 1. | | | | ❒ |
| 2. | | | | ❒ |
| 3. | | | | ❒ |
| 4. | | | | ❒ |
| 5. | | | | ❒ |

| Next Meeting Date: | Meeting Time: | Meeting Place: |
|---|---|---|
| | | |

| Title: | | |
|---|---|---|
| Duration: | Meeting Date: | Meeting Time: |
| Location: | Called By: | Type: |
| Facilitator: | Note Taker: | Time Keeper: |
| Attendees: | | |

## Agenda

| Topic: | Presenter: |
|---|---|
| 1. | |
| 2. | |
| 3. | |
| 4. | |
| 5. | |

| | Action | Person Responsible | Deadline | |
|---|---|---|---|---|
| 1. | | | | ❒ |
| 2. | | | | ❒ |
| 3. | | | | ❒ |
| 4. | | | | ❒ |
| 5. | | | | ❒ |
| 6. | | | | ❒ |
| 7. | | | | ❒ |
| 8. | | | | ❒ |
| 9. | | | | ❒ |
| 10. | | | | ❒ |

Notes:

# Agenda

| Topic: | Presenter: |
|---|---|
| 1. | |
| 2. | |
| 3. | |
| 4. | |
| 5. | |

| | Action | Person Responsible | Deadline | |
|---|---|---|---|---|
| 1. | | | | ❐ |
| 2. | | | | ❐ |
| 3. | | | | ❐ |
| 4. | | | | ❐ |
| 5. | | | | ❐ |

# Agenda

| Topic: | Presenter: |
|---|---|
| 1. | |
| 2. | |
| 3. | |
| 4. | |
| 5. | |

| | Action | Person Responsible | Deadline | |
|---|---|---|---|---|
| 1. | | | | ❐ |
| 2. | | | | ❐ |
| 3. | | | | ❐ |
| 4. | | | | ❐ |
| 5. | | | | ❐ |

| Next Meeting Date: | Meeting Time: | Meeting Place: |
|---|---|---|
| | | |

| Title: | | |
|---|---|---|
| Duration: | Meeting Date: | Meeting Time: |
| Location: | Called By: | Type: |
| Facilitator: | Note Taker: | Time Keeper: |
| Attendees: | | |

## Agenda

| Topic: | Presenter: |
|---|---|
| 1. | |
| 2. | |
| 3. | |
| 4. | |
| 5. | |

| | Action | Person Responsible | Deadline | |
|---|---|---|---|---|
| 1. | | | | ❐ |
| 2. | | | | ❐ |
| 3. | | | | ❐ |
| 4. | | | | ❐ |
| 5. | | | | ❐ |
| 6. | | | | ❐ |
| 7. | | | | ❐ |
| 8. | | | | ❐ |
| 9. | | | | ❐ |
| 10. | | | | ❐ |

Notes:

| Agenda | |
|---|---|
| Topic: | Presenter: |
| 1. | |
| 2. | |
| 3. | |
| 4. | |
| 5. | |

| | Action | Person Responsible | Deadline | |
|---|---|---|---|---|
| 1. | | | | ❐ |
| 2. | | | | ❐ |
| 3. | | | | ❐ |
| 4. | | | | ❐ |
| 5. | | | | ❐ |

| Agenda | |
|---|---|
| Topic: | Presenter: |
| 1. | |
| 2. | |
| 3. | |
| 4. | |
| 5. | |

| | Action | Person Responsible | Deadline | |
|---|---|---|---|---|
| 1. | | | | ❐ |
| 2. | | | | ❐ |
| 3. | | | | ❐ |
| 4. | | | | ❐ |
| 5. | | | | ❐ |

| Next Meeting Date: | Meeting Time: | Meeting Place: |
|---|---|---|
| | | |

| Title: | | |
|---|---|---|
| Duration: | Meeting Date: | Meeting Time: |
| Location: | Called By: | Type: |
| Facilitator: | Note Taker: | Time Keeper: |
| Attendees: | | |

## Agenda

| Topic: | Presenter: |
|---|---|
| 1. | |
| 2. | |
| 3. | |
| 4. | |
| 5. | |

| | Action | Person Responsible | Deadline | |
|---|---|---|---|---|
| 1. | | | | ❒ |
| 2. | | | | ❒ |
| 3. | | | | ❒ |
| 4. | | | | ❒ |
| 5. | | | | ❒ |
| 6. | | | | ❒ |
| 7. | | | | ❒ |
| 8. | | | | ❒ |
| 9. | | | | ❒ |
| 10. | | | | ❒ |

Notes: ______________________________

# Agenda

| Topic: | Presenter: |
|---|---|
| 1. | |
| 2. | |
| 3. | |
| 4. | |
| 5. | |

| | Action | Person Responsible | Deadline | |
|---|---|---|---|---|
| 1. | | | | ❐ |
| 2. | | | | ❐ |
| 3. | | | | ❐ |
| 4. | | | | ❐ |
| 5. | | | | ❐ |

# Agenda

| Topic: | Presenter: |
|---|---|
| 1. | |
| 2. | |
| 3. | |
| 4. | |
| 5. | |

| | Action | Person Responsible | Deadline | |
|---|---|---|---|---|
| 1. | | | | ❐ |
| 2. | | | | ❐ |
| 3. | | | | ❐ |
| 4. | | | | ❐ |
| 5. | | | | ❐ |

| Next Meeting Date: | Meeting Time: | Meeting Place: |
|---|---|---|
| | | |

| Title: | | |
|---|---|---|
| Duration: | Meeting Date: | Meeting Time: |
| Location: | Called By: | Type: |
| Facilitator: | Note Taker: | Time Keeper: |
| Attendees: | | |

## Agenda

| Topic: | Presenter: |
|---|---|
| 1. | |
| 2. | |
| 3. | |
| 4. | |
| 5. | |

| | Action | Person Responsible | Deadline | |
|---|---|---|---|---|
| 1. | | | | ❐ |
| 2. | | | | ❐ |
| 3. | | | | ❐ |
| 4. | | | | ❐ |
| 5. | | | | ❐ |
| 6. | | | | ❐ |
| 7. | | | | ❐ |
| 8. | | | | ❐ |
| 9. | | | | ❐ |
| 10. | | | | ❐ |

Notes: ______________________________

# Agenda

| Topic: | Presenter: |
|---|---|
| 1. | |
| 2. | |
| 3. | |
| 4. | |
| 5. | |

| | Action | Person Responsible | Deadline | |
|---|---|---|---|---|
| 1. | | | | ❒ |
| 2. | | | | ❒ |
| 3. | | | | ❒ |
| 4. | | | | ❒ |
| 5. | | | | ❒ |

# Agenda

| Topic: | Presenter: |
|---|---|
| 1. | |
| 2. | |
| 3. | |
| 4. | |
| 5. | |

| | Action | Person Responsible | Deadline | |
|---|---|---|---|---|
| 1. | | | | ❒ |
| 2. | | | | ❒ |
| 3. | | | | ❒ |
| 4. | | | | ❒ |
| 5. | | | | ❒ |

| Next Meeting Date: | Meeting Time: | Meeting Place: |
|---|---|---|
| | | |

| **Title:** | | |
|---|---|---|
| **Duration:** | **Meeting Date:** | **Meeting Time:** |
| **Location:** | **Called By:** | **Type:** |
| **Facilitator:** | **Note Taker:** | **Time Keeper:** |
| **Attendees:** | | |

## Agenda

| **Topic:** | **Presenter:** |
|---|---|
| 1. | |
| 2. | |
| 3. | |
| 4. | |
| 5. | |

| | Action | Person Responsible | Deadline | |
|---|---|---|---|---|
| 1. | | | | ❒ |
| 2. | | | | ❒ |
| 3. | | | | ❒ |
| 4. | | | | ❒ |
| 5. | | | | ❒ |
| 6. | | | | ❒ |
| 7. | | | | ❒ |
| 8. | | | | ❒ |
| 9. | | | | ❒ |
| 10. | | | | ❒ |

**Notes:**

## Agenda

| Topic: | Presenter: |
|---|---|
| 1. | |
| 2. | |
| 3. | |
| 4. | |
| 5. | |

| | Action | Person Responsible | Deadline | |
|---|---|---|---|---|
| 1. | | | | ❐ |
| 2. | | | | ❐ |
| 3. | | | | ❐ |
| 4. | | | | ❐ |
| 5. | | | | ❐ |

## Agenda

| Topic: | Presenter: |
|---|---|
| 1. | |
| 2. | |
| 3. | |
| 4. | |
| 5. | |

| | Action | Person Responsible | Deadline | |
|---|---|---|---|---|
| 1. | | | | ❐ |
| 2. | | | | ❐ |
| 3. | | | | ❐ |
| 4. | | | | ❐ |
| 5. | | | | ❐ |

| Next Meeting Date: | Meeting Time: | Meeting Place: |
|---|---|---|
| | | |

| **Title:** | | |
|---|---|---|
| **Duration:** | **Meeting Date:** | **Meeting Time:** |
| **Location:** | **Called By:** | **Type:** |
| **Facilitator:** | **Note Taker:** | **Time Keeper:** |
| **Attendees:** | | |

## Agenda

| **Topic:** | **Presenter:** |
|---|---|
| 1. | |
| 2. | |
| 3. | |
| 4. | |
| 5. | |

| | Action | Person Responsible | Deadline | |
|---|---|---|---|---|
| 1. | | | | ❒ |
| 2. | | | | ❒ |
| 3. | | | | ❒ |
| 4. | | | | ❒ |
| 5. | | | | ❒ |
| 6. | | | | ❒ |
| 7. | | | | ❒ |
| 8. | | | | ❒ |
| 9. | | | | ❒ |
| 10. | | | | ❒ |

**Notes:**

## Agenda

| Topic: | Presenter: |
|---|---|
| 1. | |
| 2. | |
| 3. | |
| 4. | |
| 5. | |

| | Action | Person Responsible | Deadline | |
|---|---|---|---|---|
| 1. | | | | ❐ |
| 2. | | | | ❐ |
| 3. | | | | ❐ |
| 4. | | | | ❐ |
| 5. | | | | ❐ |

## Agenda

| Topic: | Presenter: |
|---|---|
| 1. | |
| 2. | |
| 3. | |
| 4. | |
| 5. | |

| | Action | Person Responsible | Deadline | |
|---|---|---|---|---|
| 1. | | | | ❐ |
| 2. | | | | ❐ |
| 3. | | | | ❐ |
| 4. | | | | ❐ |
| 5. | | | | ❐ |

| Next Meeting Date: | Meeting Time: | Meeting Place: |
|---|---|---|
| | | |

| **Title:** | | |
|---|---|---|
| **Duration:** | **Meeting Date:** | **Meeting Time:** |
| **Location:** | **Called By:** | **Type:** |
| **Facilitator:** | **Note Taker:** | **Time Keeper:** |
| **Attendees:** | | |

## Agenda

| **Topic:** | **Presenter:** |
|---|---|
| 1. | |
| 2. | |
| 3. | |
| 4. | |
| 5. | |

| | Action | Person Responsible | Deadline | |
|---|---|---|---|---|
| 1. | | | | ❒ |
| 2. | | | | ❒ |
| 3. | | | | ❒ |
| 4. | | | | ❒ |
| 5. | | | | ❒ |
| 6. | | | | ❒ |
| 7. | | | | ❒ |
| 8. | | | | ❒ |
| 9. | | | | ❒ |
| 10. | | | | ❒ |

**Notes:**

# Agenda

| Topic: | Presenter: |
|---|---|
| 1. | |
| 2. | |
| 3. | |
| 4. | |
| 5. | |

| | Action | Person Responsible | Deadline | |
|---|---|---|---|---|
| 1. | | | | ❐ |
| 2. | | | | ❐ |
| 3. | | | | ❐ |
| 4. | | | | ❐ |
| 5. | | | | ❐ |

# Agenda

| Topic: | Presenter: |
|---|---|
| 1. | |
| 2. | |
| 3. | |
| 4. | |
| 5. | |

| | Action | Person Responsible | Deadline | |
|---|---|---|---|---|
| 1. | | | | ❐ |
| 2. | | | | ❐ |
| 3. | | | | ❐ |
| 4. | | | | ❐ |
| 5. | | | | ❐ |

| Next Meeting Date: | Meeting Time: | Meeting Place: |
|---|---|---|
| | | |

| **Title:** | | |
|---|---|---|
| **Duration:** | **Meeting Date:** | **Meeting Time:** |
| **Location:** | **Called By:** | **Type:** |
| **Facilitator:** | **Note Taker:** | **Time Keeper:** |
| **Attendees:** | | |

## Agenda

| **Topic:** | **Presenter:** |
|---|---|
| 1. | |
| 2. | |
| 3. | |
| 4. | |
| 5. | |

| | **Action** | **Person Responsible** | **Deadline** | |
|---|---|---|---|---|
| 1. | | | | ❒ |
| 2. | | | | ❒ |
| 3. | | | | ❒ |
| 4. | | | | ❒ |
| 5. | | | | ❒ |
| 6. | | | | ❒ |
| 7. | | | | ❒ |
| 8. | | | | ❒ |
| 9. | | | | ❒ |
| 10. | | | | ❒ |

**Notes:**

# Agenda

| Topic: | Presenter: |
|---|---|
| 1. | |
| 2. | |
| 3. | |
| 4. | |
| 5. | |

| | Action | Person Responsible | Deadline | |
|---|---|---|---|---|
| 1. | | | | ❒ |
| 2. | | | | ❒ |
| 3. | | | | ❒ |
| 4. | | | | ❒ |
| 5. | | | | ❒ |

# Agenda

| Topic: | Presenter: |
|---|---|
| 1. | |
| 2. | |
| 3. | |
| 4. | |
| 5. | |

| | Action | Person Responsible | Deadline | |
|---|---|---|---|---|
| 1. | | | | ❒ |
| 2. | | | | ❒ |
| 3. | | | | ❒ |
| 4. | | | | ❒ |
| 5. | | | | ❒ |

| Next Meeting Date: | Meeting Time: | Meeting Place: |
|---|---|---|
| | | |

| Title: | | |
|---|---|---|
| Duration: | Meeting Date: | Meeting Time: |
| Location: | Called By: | Type: |
| Facilitator: | Note Taker: | Time Keeper: |
| Attendees: | | |

## Agenda

| Topic: | Presenter: |
|---|---|
| 1. | |
| 2. | |
| 3. | |
| 4. | |
| 5. | |

| | Action | Person Responsible | Deadline | |
|---|---|---|---|---|
| 1. | | | | ❒ |
| 2. | | | | ❒ |
| 3. | | | | ❒ |
| 4. | | | | ❒ |
| 5. | | | | ❒ |
| 6. | | | | ❒ |
| 7. | | | | ❒ |
| 8. | | | | ❒ |
| 9. | | | | ❒ |
| 10. | | | | ❒ |

**Notes:**

# Agenda

| Topic: | Presenter: |
|---|---|
| 1. | |
| 2. | |
| 3. | |
| 4. | |
| 5. | |

| | Action | Person Responsible | Deadline | |
|---|---|---|---|---|
| 1. | | | | ❐ |
| 2. | | | | ❐ |
| 3. | | | | ❐ |
| 4. | | | | ❐ |
| 5. | | | | ❐ |

# Agenda

| Topic: | Presenter: |
|---|---|
| 1. | |
| 2. | |
| 3. | |
| 4. | |
| 5. | |

| | Action | Person Responsible | Deadline | |
|---|---|---|---|---|
| 1. | | | | ❐ |
| 2. | | | | ❐ |
| 3. | | | | ❐ |
| 4. | | | | ❐ |
| 5. | | | | ❐ |

| Next Meeting Date: | Meeting Time: | Meeting Place: |
|---|---|---|
| | | |

| Title: | | |
|---|---|---|
| Duration: | Meeting Date: | Meeting Time: |
| Location: | Called By: | Type: |
| Facilitator: | Note Taker: | Time Keeper: |
| Attendees: | | |

## Agenda

| Topic: | Presenter: |
|---|---|
| 1. | |
| 2. | |
| 3. | |
| 4. | |
| 5. | |

| | Action | Person Responsible | Deadline | |
|---|---|---|---|---|
| 1. | | | | ❐ |
| 2. | | | | ❐ |
| 3. | | | | ❐ |
| 4. | | | | ❐ |
| 5. | | | | ❐ |
| 6. | | | | ❐ |
| 7. | | | | ❐ |
| 8. | | | | ❐ |
| 9. | | | | ❐ |
| 10. | | | | ❐ |

Notes:

# Agenda

| Topic: | Presenter: |
|---|---|
| 1. | |
| 2. | |
| 3. | |
| 4. | |
| 5. | |

| | Action | Person Responsible | Deadline | |
|---|---|---|---|---|
| 1. | | | | ❐ |
| 2. | | | | ❐ |
| 3. | | | | ❐ |
| 4. | | | | ❐ |
| 5. | | | | ❐ |

# Agenda

| Topic: | Presenter: |
|---|---|
| 1. | |
| 2. | |
| 3. | |
| 4. | |
| 5. | |

| | Action | Person Responsible | Deadline | |
|---|---|---|---|---|
| 1. | | | | ❐ |
| 2. | | | | ❐ |
| 3. | | | | ❐ |
| 4. | | | | ❐ |
| 5. | | | | ❐ |

| Next Meeting Date: | Meeting Time: | Meeting Place: |
|---|---|---|
| | | |

| **Title:** | | |
|---|---|---|
| **Duration:** | **Meeting Date:** | **Meeting Time:** |
| **Location:** | **Called By:** | **Type:** |
| **Facilitator:** | **Note Taker:** | **Time Keeper:** |
| **Attendees:** | | |

## Agenda

| **Topic:** | **Presenter:** |
|---|---|
| 1. | |
| 2. | |
| 3. | |
| 4. | |
| 5. | |

| | Action | Person Responsible | Deadline | |
|---|---|---|---|---|
| 1. | | | | ❒ |
| 2. | | | | ❒ |
| 3. | | | | ❒ |
| 4. | | | | ❒ |
| 5. | | | | ❒ |
| 6. | | | | ❒ |
| 7. | | | | ❒ |
| 8. | | | | ❒ |
| 9. | | | | ❒ |
| 10. | | | | ❒ |

**Notes:**

## Agenda

| Topic: | Presenter: |
|---|---|
| 1. | |
| 2. | |
| 3. | |
| 4. | |
| 5. | |

| | Action | Person Responsible | Deadline | |
|---|---|---|---|---|
| 1. | | | | ❐ |
| 2. | | | | ❐ |
| 3. | | | | ❐ |
| 4. | | | | ❐ |
| 5. | | | | ❐ |

## Agenda

| Topic: | Presenter: |
|---|---|
| 1. | |
| 2. | |
| 3. | |
| 4. | |
| 5. | |

| | Action | Person Responsible | Deadline | |
|---|---|---|---|---|
| 1. | | | | ❐ |
| 2. | | | | ❐ |
| 3. | | | | ❐ |
| 4. | | | | ❐ |
| 5. | | | | ❐ |

| Next Meeting Date: | Meeting Time: | Meeting Place: |
|---|---|---|
| | | |

| **Title:** | | |
|---|---|---|
| **Duration:** | **Meeting Date:** | **Meeting Time:** |
| **Location:** | **Called By:** | **Type:** |
| **Facilitator:** | **Note Taker:** | **Time Keeper:** |
| **Attendees:** | | |

## Agenda

| **Topic:** | **Presenter:** |
|---|---|
| 1. | |
| 2. | |
| 3. | |
| 4. | |
| 5. | |

| | **Action** | **Person Responsible** | **Deadline** | |
|---|---|---|---|---|
| 1. | | | | ❒ |
| 2. | | | | ❒ |
| 3. | | | | ❒ |
| 4. | | | | ❒ |
| 5. | | | | ❒ |
| 6. | | | | ❒ |
| 7. | | | | ❒ |
| 8. | | | | ❒ |
| 9. | | | | ❒ |
| 10. | | | | ❒ |

**Notes:**

## Agenda

| Topic: | Presenter: |
|---|---|
| 1. | |
| 2. | |
| 3. | |
| 4. | |
| 5. | |

| | Action | Person Responsible | Deadline | |
|---|---|---|---|---|
| 1. | | | | ❐ |
| 2. | | | | ❐ |
| 3. | | | | ❐ |
| 4. | | | | ❐ |
| 5. | | | | ❐ |

## Agenda

| Topic: | Presenter: |
|---|---|
| 1. | |
| 2. | |
| 3. | |
| 4. | |
| 5. | |

| | Action | Person Responsible | Deadline | |
|---|---|---|---|---|
| 1. | | | | ❐ |
| 2. | | | | ❐ |
| 3. | | | | ❐ |
| 4. | | | | ❐ |
| 5. | | | | ❐ |

| Next Meeting Date: | Meeting Time: | Meeting Place: |
|---|---|---|
| | | |

| Title: | | |
|---|---|---|
| Duration: | Meeting Date: | Meeting Time: |
| Location: | Called By: | Type: |
| Facilitator: | Note Taker: | Time Keeper: |
| Attendees: | | |

## Agenda

| Topic: | Presenter: |
|---|---|
| 1. | |
| 2. | |
| 3. | |
| 4. | |
| 5. | |

| | Action | Person Responsible | Deadline | |
|---|---|---|---|---|
| 1. | | | | ❐ |
| 2. | | | | ❐ |
| 3. | | | | ❐ |
| 4. | | | | ❐ |
| 5. | | | | ❐ |
| 6. | | | | ❐ |
| 7. | | | | ❐ |
| 8. | | | | ❐ |
| 9. | | | | ❐ |
| 10. | | | | ❐ |

Notes:

# Agenda

| Topic: | Presenter: |
|---|---|
| 1. | |
| 2. | |
| 3. | |
| 4. | |
| 5. | |

| | Action | Person Responsible | Deadline | |
|---|---|---|---|---|
| 1. | | | | ❐ |
| 2. | | | | ❐ |
| 3. | | | | ❐ |
| 4. | | | | ❐ |
| 5. | | | | ❐ |

# Agenda

| Topic: | Presenter: |
|---|---|
| 1. | |
| 2. | |
| 3. | |
| 4. | |
| 5. | |

| | Action | Person Responsible | Deadline | |
|---|---|---|---|---|
| 1. | | | | ❐ |
| 2. | | | | ❐ |
| 3. | | | | ❐ |
| 4. | | | | ❐ |
| 5. | | | | ❐ |

| Next Meeting Date: | Meeting Time: | Meeting Place: |
|---|---|---|
| | | |

| Title: | | |
|---|---|---|
| Duration: | Meeting Date: | Meeting Time: |
| Location: | Called By: | Type: |
| Facilitator: | Note Taker: | Time Keeper: |
| Attendees: | | |

## Agenda

| Topic: | Presenter: |
|---|---|
| 1. | |
| 2. | |
| 3. | |
| 4. | |
| 5. | |

| | Action | Person Responsible | Deadline | |
|---|---|---|---|---|
| 1. | | | | ❐ |
| 2. | | | | ❐ |
| 3. | | | | ❐ |
| 4. | | | | ❐ |
| 5. | | | | ❐ |
| 6. | | | | ❐ |
| 7. | | | | ❐ |
| 8. | | | | ❐ |
| 9. | | | | ❐ |
| 10. | | | | ❐ |

**Notes:**

## Agenda

| Topic: | Presenter: |
|---|---|
| 1. | |
| 2. | |
| 3. | |
| 4. | |
| 5. | |

| | Action | Person Responsible | Deadline | |
|---|---|---|---|---|
| 1. | | | | ❐ |
| 2. | | | | ❐ |
| 3. | | | | ❐ |
| 4. | | | | ❐ |
| 5. | | | | ❐ |

## Agenda

| Topic: | Presenter: |
|---|---|
| 1. | |
| 2. | |
| 3. | |
| 4. | |
| 5. | |

| | Action | Person Responsible | Deadline | |
|---|---|---|---|---|
| 1. | | | | ❐ |
| 2. | | | | ❐ |
| 3. | | | | ❐ |
| 4. | | | | ❐ |
| 5. | | | | ❐ |

| Next Meeting Date: | Meeting Time: | Meeting Place: |
|---|---|---|
| | | |

| **Title:** | | |
|---|---|---|
| **Duration:** | **Meeting Date:** | **Meeting Time:** |
| **Location:** | **Called By:** | **Type:** |
| **Facilitator:** | **Note Taker:** | **Time Keeper:** |
| **Attendees:** | | |

## Agenda

| **Topic:** | **Presenter:** |
|---|---|
| 1. | |
| 2. | |
| 3. | |
| 4. | |
| 5. | |

| | **Action** | **Person Responsible** | **Deadline** | |
|---|---|---|---|---|
| 1. | | | | ❐ |
| 2. | | | | ❐ |
| 3. | | | | ❐ |
| 4. | | | | ❐ |
| 5. | | | | ❐ |
| 6. | | | | ❐ |
| 7. | | | | ❐ |
| 8. | | | | ❐ |
| 9. | | | | ❐ |
| 10. | | | | ❐ |

**Notes:**

## Agenda

| Topic: | Presenter: |
|---|---|
| 1. | |
| 2. | |
| 3. | |
| 4. | |
| 5. | |

| | Action | Person Responsible | Deadline | |
|---|---|---|---|---|
| 1. | | | | ❒ |
| 2. | | | | ❒ |
| 3. | | | | ❒ |
| 4. | | | | ❒ |
| 5. | | | | ❒ |

## Agenda

| Topic: | Presenter: |
|---|---|
| 1. | |
| 2. | |
| 3. | |
| 4. | |
| 5. | |

| | Action | Person Responsible | Deadline | |
|---|---|---|---|---|
| 1. | | | | ❒ |
| 2. | | | | ❒ |
| 3. | | | | ❒ |
| 4. | | | | ❒ |
| 5. | | | | ❒ |

| Next Meeting Date: | Meeting Time: | Meeting Place: |
|---|---|---|
| | | |

| **Title:** | | |
|---|---|---|
| **Duration:** | **Meeting Date:** | **Meeting Time:** |
| **Location:** | **Called By:** | **Type:** |
| **Facilitator:** | **Note Taker:** | **Time Keeper:** |
| **Attendees:** | | |

| Agenda | |
|---|---|
| **Topic:** | **Presenter:** |
| 1. | |
| 2. | |
| 3. | |
| 4. | |
| 5. | |

| | Action | Person Responsible | Deadline | |
|---|---|---|---|---|
| 1. | | | | ❒ |
| 2. | | | | ❒ |
| 3. | | | | ❒ |
| 4. | | | | ❒ |
| 5. | | | | ❒ |
| 6. | | | | ❒ |
| 7. | | | | ❒ |
| 8. | | | | ❒ |
| 9. | | | | ❒ |
| 10. | | | | ❒ |

**Notes:**

# Agenda

| Topic: | Presenter: |
|---|---|
| 1. | |
| 2. | |
| 3. | |
| 4. | |
| 5. | |

| | Action | Person Responsible | Deadline | |
|---|---|---|---|---|
| 1. | | | | ❐ |
| 2. | | | | ❐ |
| 3. | | | | ❐ |
| 4. | | | | ❐ |
| 5. | | | | ❐ |

# Agenda

| Topic: | Presenter: |
|---|---|
| 1. | |
| 2. | |
| 3. | |
| 4. | |
| 5. | |

| | Action | Person Responsible | Deadline | |
|---|---|---|---|---|
| 1. | | | | ❐ |
| 2. | | | | ❐ |
| 3. | | | | ❐ |
| 4. | | | | ❐ |
| 5. | | | | ❐ |

| Next Meeting Date: | Meeting Time: | Meeting Place: |
|---|---|---|
| | | |

| Title: | | |
|---|---|---|
| Duration: | Meeting Date: | Meeting Time: |
| Location: | Called By: | Type: |
| Facilitator: | Note Taker: | Time Keeper: |
| Attendees: | | |

## Agenda

| Topic: | Presenter: |
|---|---|
| 1. | |
| 2. | |
| 3. | |
| 4. | |
| 5. | |

| | Action | Person Responsible | Deadline | |
|---|---|---|---|---|
| 1. | | | | ❐ |
| 2. | | | | ❐ |
| 3. | | | | ❐ |
| 4. | | | | ❐ |
| 5. | | | | ❐ |
| 6. | | | | ❐ |
| 7. | | | | ❐ |
| 8. | | | | ❐ |
| 9. | | | | ❐ |
| 10. | | | | ❐ |

**Notes:**

# Agenda

| Topic: | Presenter: |
|---|---|
| 1. | |
| 2. | |
| 3. | |
| 4. | |
| 5. | |

| | Action | Person Responsible | Deadline | |
|---|---|---|---|---|
| 1. | | | | ❐ |
| 2. | | | | ❐ |
| 3. | | | | ❐ |
| 4. | | | | ❐ |
| 5. | | | | ❐ |

# Agenda

| Topic: | Presenter: |
|---|---|
| 1. | |
| 2. | |
| 3. | |
| 4. | |
| 5. | |

| | Action | Person Responsible | Deadline | |
|---|---|---|---|---|
| 1. | | | | ❐ |
| 2. | | | | ❐ |
| 3. | | | | ❐ |
| 4. | | | | ❐ |
| 5. | | | | ❐ |

| Next Meeting Date: | Meeting Time: | Meeting Place: |
|---|---|---|
| | | |

| Title: | | |
|---|---|---|
| Duration: | Meeting Date: | Meeting Time: |
| Location: | Called By: | Type: |
| Facilitator: | Note Taker: | Time Keeper: |
| Attendees: | | |

## Agenda

| Topic: | Presenter: |
|---|---|
| 1. | |
| 2. | |
| 3. | |
| 4. | |
| 5. | |

| | Action | Person Responsible | Deadline | |
|---|---|---|---|---|
| 1. | | | | ❐ |
| 2. | | | | ❐ |
| 3. | | | | ❐ |
| 4. | | | | ❐ |
| 5. | | | | ❐ |
| 6. | | | | ❐ |
| 7. | | | | ❐ |
| 8. | | | | ❐ |
| 9. | | | | ❐ |
| 10. | | | | ❐ |

Notes:

| Agenda | |
|---|---|
| **Topic:** | **Presenter:** |
| 1. | |
| 2. | |
| 3. | |
| 4. | |
| 5. | |

| | Action | Person Responsible | Deadline | |
|---|---|---|---|---|
| 1. | | | | ❐ |
| 2. | | | | ❐ |
| 3. | | | | ❐ |
| 4. | | | | ❐ |
| 5. | | | | ❐ |

| Agenda | |
|---|---|
| **Topic:** | **Presenter:** |
| 1. | |
| 2. | |
| 3. | |
| 4. | |
| 5. | |

| | Action | Person Responsible | Deadline | |
|---|---|---|---|---|
| 1. | | | | ❐ |
| 2. | | | | ❐ |
| 3. | | | | ❐ |
| 4. | | | | ❐ |
| 5. | | | | ❐ |

| Next Meeting Date: | Meeting Time: | Meeting Place: |
|---|---|---|
| | | |

| Title: | | |
|---|---|---|
| Duration: | Meeting Date: | Meeting Time: |
| Location: | Called By: | Type: |
| Facilitator: | Note Taker: | Time Keeper: |
| Attendees: | | |

## Agenda

| Topic: | Presenter: |
|---|---|
| 1. | |
| 2. | |
| 3. | |
| 4. | |
| 5. | |

| | Action | Person Responsible | Deadline | |
|---|---|---|---|---|
| 1. | | | | ❐ |
| 2. | | | | ❐ |
| 3. | | | | ❐ |
| 4. | | | | ❐ |
| 5. | | | | ❐ |
| 6. | | | | ❐ |
| 7. | | | | ❐ |
| 8. | | | | ❐ |
| 9. | | | | ❐ |
| 10. | | | | ❐ |

Notes:

## Agenda

| Topic: | Presenter: |
|---|---|
| 1. | |
| 2. | |
| 3. | |
| 4. | |
| 5. | |

| | Action | Person Responsible | Deadline | |
|---|---|---|---|---|
| 1. | | | | ❒ |
| 2. | | | | ❒ |
| 3. | | | | ❒ |
| 4. | | | | ❒ |
| 5. | | | | ❒ |

## Agenda

| Topic: | Presenter: |
|---|---|
| 1. | |
| 2. | |
| 3. | |
| 4. | |
| 5. | |

| | Action | Person Responsible | Deadline | |
|---|---|---|---|---|
| 1. | | | | ❒ |
| 2. | | | | ❒ |
| 3. | | | | ❒ |
| 4. | | | | ❒ |
| 5. | | | | ❒ |

| Next Meeting Date: | Meeting Time: | Meeting Place: |
|---|---|---|
| | | |

| **Title:** | | |
|---|---|---|
| **Duration:** | **Meeting Date:** | **Meeting Time:** |
| **Location:** | **Called By:** | **Type:** |
| **Facilitator:** | **Note Taker:** | **Time Keeper:** |
| **Attendees:** | | |

## Agenda

| **Topic:** | **Presenter:** |
|---|---|
| 1. | |
| 2. | |
| 3. | |
| 4. | |
| 5. | |

| | **Action** | **Person Responsible** | **Deadline** | |
|---|---|---|---|---|
| 1. | | | | ❐ |
| 2. | | | | ❐ |
| 3. | | | | ❐ |
| 4. | | | | ❐ |
| 5. | | | | ❐ |
| 6. | | | | ❐ |
| 7. | | | | ❐ |
| 8. | | | | ❐ |
| 9. | | | | ❐ |
| 10. | | | | ❐ |

**Notes:**

## Agenda

| Topic: | Presenter: |
|---|---|
| 1. | |
| 2. | |
| 3. | |
| 4. | |
| 5. | |

| | Action | Person Responsible | Deadline | |
|---|---|---|---|---|
| 1. | | | | ❒ |
| 2. | | | | ❒ |
| 3. | | | | ❒ |
| 4. | | | | ❒ |
| 5. | | | | ❒ |

## Agenda

| Topic: | Presenter: |
|---|---|
| 1. | |
| 2. | |
| 3. | |
| 4. | |
| 5. | |

| | Action | Person Responsible | Deadline | |
|---|---|---|---|---|
| 1. | | | | ❒ |
| 2. | | | | ❒ |
| 3. | | | | ❒ |
| 4. | | | | ❒ |
| 5. | | | | ❒ |

| Next Meeting Date: | Meeting Time: | Meeting Place: |
|---|---|---|
| | | |

| Title: | | |
|---|---|---|
| Duration: | Meeting Date: | Meeting Time: |
| Location: | Called By: | Type: |
| Facilitator: | Note Taker: | Time Keeper: |
| Attendees: | | |

## Agenda

| Topic: | Presenter: |
|---|---|
| 1. | |
| 2. | |
| 3. | |
| 4. | |
| 5. | |

| | Action | Person Responsible | Deadline | |
|---|---|---|---|---|
| 1. | | | | ❐ |
| 2. | | | | ❐ |
| 3. | | | | ❐ |
| 4. | | | | ❐ |
| 5. | | | | ❐ |
| 6. | | | | ❐ |
| 7. | | | | ❐ |
| 8. | | | | ❐ |
| 9. | | | | ❐ |
| 10. | | | | ❐ |

**Notes:**

## Agenda

| Topic: | Presenter: |
| --- | --- |

1.
2.
3.
4.
5.

| | Action | Person Responsible | Deadline | |
| --- | --- | --- | --- | --- |
| 1. | | | | ❐ |
| 2. | | | | ❐ |
| 3. | | | | ❐ |
| 4. | | | | ❐ |
| 5. | | | | ❐ |

## Agenda

| Topic: | Presenter: |
| --- | --- |

1.
2.
3.
4.
5.

| | Action | Person Responsible | Deadline | |
| --- | --- | --- | --- | --- |
| 1. | | | | ❐ |
| 2. | | | | ❐ |
| 3. | | | | ❐ |
| 4. | | | | ❐ |
| 5. | | | | ❐ |

| Next Meeting Date: | Meeting Time: | Meeting Place: |
| --- | --- | --- |

| Title: | | |
|---|---|---|
| Duration: | Meeting Date: | Meeting Time: |
| Location: | Called By: | Type: |
| Facilitator: | Note Taker: | Time Keeper: |
| Attendees: | | |

## Agenda

| Topic: | Presenter: |
|---|---|
| 1. | |
| 2. | |
| 3. | |
| 4. | |
| 5. | |

| | Action | Person Responsible | Deadline | |
|---|---|---|---|---|
| 1. | | | | ❐ |
| 2. | | | | ❐ |
| 3. | | | | ❐ |
| 4. | | | | ❐ |
| 5. | | | | ❐ |
| 6. | | | | ❐ |
| 7. | | | | ❐ |
| 8. | | | | ❐ |
| 9. | | | | ❐ |
| 10. | | | | ❐ |

Notes:

# Agenda

| Topic: | Presenter: |
|---|---|
| 1. | |
| 2. | |
| 3. | |
| 4. | |
| 5. | |

| | Action | Person Responsible | Deadline | |
|---|---|---|---|---|
| 1. | | | | ❒ |
| 2. | | | | ❒ |
| 3. | | | | ❒ |
| 4. | | | | ❒ |
| 5. | | | | ❒ |

# Agenda

| Topic: | Presenter: |
|---|---|
| 1. | |
| 2. | |
| 3. | |
| 4. | |
| 5. | |

| | Action | Person Responsible | Deadline | |
|---|---|---|---|---|
| 1. | | | | ❒ |
| 2. | | | | ❒ |
| 3. | | | | ❒ |
| 4. | | | | ❒ |
| 5. | | | | ❒ |

| Next Meeting Date: | Meeting Time: | Meeting Place: |
|---|---|---|
| | | |

| **Title:** | | |
|---|---|---|
| **Duration:** | **Meeting Date:** | **Meeting Time:** |
| **Location:** | **Called By:** | **Type:** |
| **Facilitator:** | **Note Taker:** | **Time Keeper:** |
| **Attendees:** | | |

## Agenda

| **Topic:** | **Presenter:** |
|---|---|
| 1. | |
| 2. | |
| 3. | |
| 4. | |
| 5. | |

| | Action | Person Responsible | Deadline | |
|---|---|---|---|---|
| 1. | | | | ❒ |
| 2. | | | | ❒ |
| 3. | | | | ❒ |
| 4. | | | | ❒ |
| 5. | | | | ❒ |
| 6. | | | | ❒ |
| 7. | | | | ❒ |
| 8. | | | | ❒ |
| 9. | | | | ❒ |
| 10. | | | | ❒ |

**Notes:**

## Agenda

| Topic: | Presenter: |
| --- | --- |
| 1. | |
| 2. | |
| 3. | |
| 4. | |
| 5. | |

| | Action | Person Responsible | Deadline | |
| --- | --- | --- | --- | --- |
| 1. | | | | ❒ |
| 2. | | | | ❒ |
| 3. | | | | ❒ |
| 4. | | | | ❒ |
| 5. | | | | ❒ |

## Agenda

| Topic: | Presenter: |
| --- | --- |
| 1. | |
| 2. | |
| 3. | |
| 4. | |
| 5. | |

| | Action | Person Responsible | Deadline | |
| --- | --- | --- | --- | --- |
| 1. | | | | ❒ |
| 2. | | | | ❒ |
| 3. | | | | ❒ |
| 4. | | | | ❒ |
| 5. | | | | ❒ |

| Next Meeting Date: | Meeting Time: | Meeting Place: |
| --- | --- | --- |

| **Title:** | | |
|---|---|---|
| **Duration:** | **Meeting Date:** | **Meeting Time:** |
| **Location:** | **Called By:** | **Type:** |
| **Facilitator:** | **Note Taker:** | **Time Keeper:** |
| **Attendees:** | | |

## Agenda

| **Topic:** | **Presenter:** |
|---|---|
| 1. | |
| 2. | |
| 3. | |
| 4. | |
| 5. | |

| | Action | Person Responsible | Deadline | |
|---|---|---|---|---|
| 1. | | | | ❐ |
| 2. | | | | ❐ |
| 3. | | | | ❐ |
| 4. | | | | ❐ |
| 5. | | | | ❐ |
| 6. | | | | ❐ |
| 7. | | | | ❐ |
| 8. | | | | ❐ |
| 9. | | | | ❐ |
| 10. | | | | ❐ |

**Notes:** ______________________________________________

______________________________________________

______________________________________________

______________________________________________

______________________________________________

______________________________________________

______________________________________________

| Agenda | |
|---|---|
| **Topic:** | **Presenter:** |
| 1. | |
| 2. | |
| 3. | |
| 4. | |
| 5. | |

| | Action | Person Responsible | Deadline | |
|---|---|---|---|---|
| 1. | | | | ❐ |
| 2. | | | | ❐ |
| 3. | | | | ❐ |
| 4. | | | | ❐ |
| 5. | | | | ❐ |

| Agenda | |
|---|---|
| **Topic:** | **Presenter:** |
| 1. | |
| 2. | |
| 3. | |
| 4. | |
| 5. | |

| | Action | Person Responsible | Deadline | |
|---|---|---|---|---|
| 1. | | | | ❐ |
| 2. | | | | ❐ |
| 3. | | | | ❐ |
| 4. | | | | ❐ |
| 5. | | | | ❐ |

| Next Meeting Date: | Meeting Time: | Meeting Place: |
|---|---|---|
| | | |

| Title: | | |
|---|---|---|
| Duration: | Meeting Date: | Meeting Time: |
| Location: | Called By: | Type: |
| Facilitator: | Note Taker: | Time Keeper: |
| Attendees: | | |

| Agenda | |
|---|---|
| Topic: | Presenter: |
| 1. | |
| 2. | |
| 3. | |
| 4. | |
| 5. | |

| | Action | Person Responsible | Deadline | |
|---|---|---|---|---|
| 1. | | | | ❒ |
| 2. | | | | ❒ |
| 3. | | | | ❒ |
| 4. | | | | ❒ |
| 5. | | | | ❒ |
| 6. | | | | ❒ |
| 7. | | | | ❒ |
| 8. | | | | ❒ |
| 9. | | | | ❒ |
| 10. | | | | ❒ |

Notes:

## Agenda

| Topic: | Presenter: |
|---|---|
| 1. | |
| 2. | |
| 3. | |
| 4. | |
| 5. | |

| | Action | Person Responsible | Deadline | |
|---|---|---|---|---|
| 1. | | | | ❒ |
| 2. | | | | ❒ |
| 3. | | | | ❒ |
| 4. | | | | ❒ |
| 5. | | | | ❒ |

## Agenda

| Topic: | Presenter: |
|---|---|
| 1. | |
| 2. | |
| 3. | |
| 4. | |
| 5. | |

| | Action | Person Responsible | Deadline | |
|---|---|---|---|---|
| 1. | | | | ❒ |
| 2. | | | | ❒ |
| 3. | | | | ❒ |
| 4. | | | | ❒ |
| 5. | | | | ❒ |

| Next Meeting Date: | Meeting Time: | Meeting Place: |
|---|---|---|
| | | |

| Title: | | |
|---|---|---|
| Duration: | Meeting Date: | Meeting Time: |
| Location: | Called By: | Type: |
| Facilitator: | Note Taker: | Time Keeper: |
| Attendees: | | |

## Agenda

| Topic: | Presenter: |
|---|---|
| 1. | |
| 2. | |
| 3. | |
| 4. | |
| 5. | |

| | Action | Person Responsible | Deadline | |
|---|---|---|---|---|
| 1. | | | | ❒ |
| 2. | | | | ❒ |
| 3. | | | | ❒ |
| 4. | | | | ❒ |
| 5. | | | | ❒ |
| 6. | | | | ❒ |
| 7. | | | | ❒ |
| 8. | | | | ❒ |
| 9. | | | | ❒ |
| 10. | | | | ❒ |

Notes:

## Agenda

| Topic: | Presenter: |
|---|---|
| 1. | |
| 2. | |
| 3. | |
| 4. | |
| 5. | |

| | Action | Person Responsible | Deadline | |
|---|---|---|---|---|
| 1. | | | | ❒ |
| 2. | | | | ❒ |
| 3. | | | | ❒ |
| 4. | | | | ❒ |
| 5. | | | | ❒ |

## Agenda

| Topic: | Presenter: |
|---|---|
| 1. | |
| 2. | |
| 3. | |
| 4. | |
| 5. | |

| | Action | Person Responsible | Deadline | |
|---|---|---|---|---|
| 1. | | | | ❒ |
| 2. | | | | ❒ |
| 3. | | | | ❒ |
| 4. | | | | ❒ |
| 5. | | | | ❒ |

| Next Meeting Date: | Meeting Time: | Meeting Place: |
|---|---|---|
| | | |

| **Title:** | | |
|---|---|---|
| **Duration:** | **Meeting Date:** | **Meeting Time:** |
| **Location:** | **Called By:** | **Type:** |
| **Facilitator:** | **Note Taker:** | **Time Keeper:** |
| **Attendees:** | | |

# Agenda

| **Topic:** | **Presenter:** |
|---|---|
| 1. | |
| 2. | |
| 3. | |
| 4. | |
| 5. | |

| | **Action** | **Person Responsible** | **Deadline** | |
|---|---|---|---|---|
| 1. | | | | ❐ |
| 2. | | | | ❐ |
| 3. | | | | ❐ |
| 4. | | | | ❐ |
| 5. | | | | ❐ |
| 6. | | | | ❐ |
| 7. | | | | ❐ |
| 8. | | | | ❐ |
| 9. | | | | ❐ |
| 10. | | | | ❐ |

**Notes:**

# Agenda

| Topic: | Presenter: |
|---|---|
| 1. | |
| 2. | |
| 3. | |
| 4. | |
| 5. | |

| | Action | Person Responsible | Deadline | |
|---|---|---|---|---|
| 1. | | | | ❒ |
| 2. | | | | ❒ |
| 3. | | | | ❒ |
| 4. | | | | ❒ |
| 5. | | | | ❒ |

# Agenda

| Topic: | Presenter: |
|---|---|
| 1. | |
| 2. | |
| 3. | |
| 4. | |
| 5. | |

| | Action | Person Responsible | Deadline | |
|---|---|---|---|---|
| 1. | | | | ❒ |
| 2. | | | | ❒ |
| 3. | | | | ❒ |
| 4. | | | | ❒ |
| 5. | | | | ❒ |

| Next Meeting Date: | Meeting Time: | Meeting Place: |
|---|---|---|
| | | |

| **Title:** | | |
|---|---|---|
| **Duration:** | **Meeting Date:** | **Meeting Time:** |
| **Location:** | **Called By:** | **Type:** |
| **Facilitator:** | **Note Taker:** | **Time Keeper:** |
| **Attendees:** | | |

## Agenda

| **Topic:** | **Presenter:** |
|---|---|
| 1. | |
| 2. | |
| 3. | |
| 4. | |
| 5. | |

| | Action | Person Responsible | Deadline | |
|---|---|---|---|---|
| 1. | | | | ❐ |
| 2. | | | | ❐ |
| 3. | | | | ❐ |
| 4. | | | | ❐ |
| 5. | | | | ❐ |
| 6. | | | | ❐ |
| 7. | | | | ❐ |
| 8. | | | | ❐ |
| 9. | | | | ❐ |
| 10. | | | | ❐ |

**Notes:**

# Agenda

| Topic: | Presenter: |
|---|---|
| 1. | |
| 2. | |
| 3. | |
| 4. | |
| 5. | |

| | Action | Person Responsible | Deadline | |
|---|---|---|---|---|
| 1. | | | | ❒ |
| 2. | | | | ❒ |
| 3. | | | | ❒ |
| 4. | | | | ❒ |
| 5. | | | | ❒ |

# Agenda

| Topic: | Presenter: |
|---|---|
| 1. | |
| 2. | |
| 3. | |
| 4. | |
| 5. | |

| | Action | Person Responsible | Deadline | |
|---|---|---|---|---|
| 1. | | | | ❒ |
| 2. | | | | ❒ |
| 3. | | | | ❒ |
| 4. | | | | ❒ |
| 5. | | | | ❒ |

| Next Meeting Date: | Meeting Time: | Meeting Place: |
|---|---|---|
| | | |

| **Title:** | | |
|---|---|---|
| **Duration:** | **Meeting Date:** | **Meeting Time:** |
| **Location:** | **Called By:** | **Type:** |
| **Facilitator:** | **Note Taker:** | **Time Keeper:** |
| **Attendees:** | | |

| **Agenda** | |
|---|---|
| **Topic:** | **Presenter:** |
| 1. | |
| 2. | |
| 3. | |
| 4. | |
| 5. | |

| | Action | Person Responsible | Deadline | |
|---|---|---|---|---|
| 1. | | | | ❐ |
| 2. | | | | ❐ |
| 3. | | | | ❐ |
| 4. | | | | ❐ |
| 5. | | | | ❐ |
| 6. | | | | ❐ |
| 7. | | | | ❐ |
| 8. | | | | ❐ |
| 9. | | | | ❐ |
| 10. | | | | ❐ |

**Notes:**

# Agenda

| Topic: | Presenter: |
|---|---|
| 1. | |
| 2. | |
| 3. | |
| 4. | |
| 5. | |

| | Action | Person Responsible | Deadline | |
|---|---|---|---|---|
| 1. | | | | ❐ |
| 2. | | | | ❐ |
| 3. | | | | ❐ |
| 4. | | | | ❐ |
| 5. | | | | ❐ |

# Agenda

| Topic: | Presenter: |
|---|---|
| 1. | |
| 2. | |
| 3. | |
| 4. | |
| 5. | |

| | Action | Person Responsible | Deadline | |
|---|---|---|---|---|
| 1. | | | | ❐ |
| 2. | | | | ❐ |
| 3. | | | | ❐ |
| 4. | | | | ❐ |
| 5. | | | | ❐ |

| Next Meeting Date: | Meeting Time: | Meeting Place: |
|---|---|---|
| | | |

| **Title:** | | |
|---|---|---|
| **Duration:** | **Meeting Date:** | **Meeting Time:** |
| **Location:** | **Called By:** | **Type:** |
| **Facilitator:** | **Note Taker:** | **Time Keeper:** |
| **Attendees:** | | |

## Agenda

| **Topic:** | **Presenter:** |
|---|---|
| 1. | |
| 2. | |
| 3. | |
| 4. | |
| 5. | |

| | Action | Person Responsible | Deadline | |
|---|---|---|---|---|
| 1. | | | | ❒ |
| 2. | | | | ❒ |
| 3. | | | | ❒ |
| 4. | | | | ❒ |
| 5. | | | | ❒ |
| 6. | | | | ❒ |
| 7. | | | | ❒ |
| 8. | | | | ❒ |
| 9. | | | | ❒ |
| 10. | | | | ❒ |

**Notes:**

| Agenda | |
|---|---|
| **Topic:** | **Presenter:** |
| 1. | |
| 2. | |
| 3. | |
| 4. | |
| 5. | |

| | Action | Person Responsible | Deadline | |
|---|---|---|---|---|
| 1. | | | | ❐ |
| 2. | | | | ❐ |
| 3. | | | | ❐ |
| 4. | | | | ❐ |
| 5. | | | | ❐ |

| Agenda | |
|---|---|
| **Topic:** | **Presenter:** |
| 1. | |
| 2. | |
| 3. | |
| 4. | |
| 5. | |

| | Action | Person Responsible | Deadline | |
|---|---|---|---|---|
| 1. | | | | ❐ |
| 2. | | | | ❐ |
| 3. | | | | ❐ |
| 4. | | | | ❐ |
| 5. | | | | ❐ |

| Next Meeting Date: | Meeting Time: | Meeting Place: |
|---|---|---|
| | | |

| **Title:** | | |
|---|---|---|
| **Duration:** | **Meeting Date:** | **Meeting Time:** |
| **Location:** | **Called By:** | **Type:** |
| **Facilitator:** | **Note Taker:** | **Time Keeper:** |
| **Attendees:** | | |

## Agenda

| **Topic:** | **Presenter:** |
|---|---|
| 1. | |
| 2. | |
| 3. | |
| 4. | |
| 5. | |

| | Action | Person Responsible | Deadline | |
|---|---|---|---|---|
| 1. | | | | ❒ |
| 2. | | | | ❒ |
| 3. | | | | ❒ |
| 4. | | | | ❒ |
| 5. | | | | ❒ |
| 6. | | | | ❒ |
| 7. | | | | ❒ |
| 8. | | | | ❒ |
| 9. | | | | ❒ |
| 10. | | | | ❒ |

**Notes:**

# Agenda

| Topic: | Presenter: |
|---|---|
| 1. | |
| 2. | |
| 3. | |
| 4. | |
| 5. | |

| | Action | Person Responsible | Deadline | |
|---|---|---|---|---|
| 1. | | | | ❒ |
| 2. | | | | ❒ |
| 3. | | | | ❒ |
| 4. | | | | ❒ |
| 5. | | | | ❒ |

# Agenda

| Topic: | Presenter: |
|---|---|
| 1. | |
| 2. | |
| 3. | |
| 4. | |
| 5. | |

| | Action | Person Responsible | Deadline | |
|---|---|---|---|---|
| 1. | | | | ❒ |
| 2. | | | | ❒ |
| 3. | | | | ❒ |
| 4. | | | | ❒ |
| 5. | | | | ❒ |

| Next Meeting Date: | Meeting Time: | Meeting Place: |
|---|---|---|
| | | |

| **Title:** | | |
|---|---|---|
| **Duration:** | **Meeting Date:** | **Meeting Time:** |
| **Location:** | **Called By:** | **Type:** |
| **Facilitator:** | **Note Taker:** | **Time Keeper:** |
| **Attendees:** | | |

| Agenda | |
|---|---|
| **Topic:** | **Presenter:** |
| 1. | |
| 2. | |
| 3. | |
| 4. | |
| 5. | |

| | Action | Person Responsible | Deadline | |
|---|---|---|---|---|
| 1. | | | | ❒ |
| 2. | | | | ❒ |
| 3. | | | | ❒ |
| 4. | | | | ❒ |
| 5. | | | | ❒ |
| 6. | | | | ❒ |
| 7. | | | | ❒ |
| 8. | | | | ❒ |
| 9. | | | | ❒ |
| 10. | | | | ❒ |

**Notes:**

# Agenda

| Topic: | Presenter: |
|---|---|
| 1. | |
| 2. | |
| 3. | |
| 4. | |
| 5. | |

| | Action | Person Responsible | Deadline | |
|---|---|---|---|---|
| 1. | | | | ❒ |
| 2. | | | | ❒ |
| 3. | | | | ❒ |
| 4. | | | | ❒ |
| 5. | | | | ❒ |

# Agenda

| Topic: | Presenter: |
|---|---|
| 1. | |
| 2. | |
| 3. | |
| 4. | |
| 5. | |

| | Action | Person Responsible | Deadline | |
|---|---|---|---|---|
| 1. | | | | ❒ |
| 2. | | | | ❒ |
| 3. | | | | ❒ |
| 4. | | | | ❒ |
| 5. | | | | ❒ |

| Next Meeting Date: | Meeting Time: | Meeting Place: |
|---|---|---|
| | | |

| **Title:** | | |
|---|---|---|
| **Duration:** | **Meeting Date:** | **Meeting Time:** |
| **Location:** | **Called By:** | **Type:** |
| **Facilitator:** | **Note Taker:** | **Time Keeper:** |
| **Attendees:** | | |

| Agenda | |
|---|---|
| **Topic:** | **Presenter:** |
| 1. | |
| 2. | |
| 3. | |
| 4. | |
| 5. | |

| | Action | Person Responsible | Deadline | |
|---|---|---|---|---|
| 1. | | | | ❐ |
| 2. | | | | ❐ |
| 3. | | | | ❐ |
| 4. | | | | ❐ |
| 5. | | | | ❐ |
| 6. | | | | ❐ |
| 7. | | | | ❐ |
| 8. | | | | ❐ |
| 9. | | | | ❐ |
| 10. | | | | ❐ |

**Notes:**

# Agenda

| Topic: | Presenter: |
|---|---|
| 1. | |
| 2. | |
| 3. | |
| 4. | |
| 5. | |

| | Action | Person Responsible | Deadline | |
|---|---|---|---|---|
| 1. | | | | ❒ |
| 2. | | | | ❒ |
| 3. | | | | ❒ |
| 4. | | | | ❒ |
| 5. | | | | ❒ |

# Agenda

| Topic: | Presenter: |
|---|---|
| 1. | |
| 2. | |
| 3. | |
| 4. | |
| 5. | |

| | Action | Person Responsible | Deadline | |
|---|---|---|---|---|
| 1. | | | | ❒ |
| 2. | | | | ❒ |
| 3. | | | | ❒ |
| 4. | | | | ❒ |
| 5. | | | | ❒ |

| Next Meeting Date: | Meeting Time: | Meeting Place: |
|---|---|---|
| | | |

| Title: | | |
|---|---|---|
| Duration: | Meeting Date: | Meeting Time: |
| Location: | Called By: | Type: |
| Facilitator: | Note Taker: | Time Keeper: |
| Attendees: | | |

| Agenda | |
|---|---|
| Topic: | Presenter: |
| 1. | |
| 2. | |
| 3. | |
| 4. | |
| 5. | |

| | Action | Person Responsible | Deadline | |
|---|---|---|---|---|
| 1. | | | | ❐ |
| 2. | | | | ❐ |
| 3. | | | | ❐ |
| 4. | | | | ❐ |
| 5. | | | | ❐ |
| 6. | | | | ❐ |
| 7. | | | | ❐ |
| 8. | | | | ❐ |
| 9. | | | | ❐ |
| 10. | | | | ❐ |

Notes:

| Agenda | |
|---|---|
| Topic: | Presenter: |
| 1. | |
| 2. | |
| 3. | |
| 4. | |
| 5. | |

| | Action | Person Responsible | Deadline | |
|---|---|---|---|---|
| 1. | | | | ❐ |
| 2. | | | | ❐ |
| 3. | | | | ❐ |
| 4. | | | | ❐ |
| 5. | | | | ❐ |

| Agenda | |
|---|---|
| Topic: | Presenter: |
| 1. | |
| 2. | |
| 3. | |
| 4. | |
| 5. | |

| | Action | Person Responsible | Deadline | |
|---|---|---|---|---|
| 1. | | | | ❐ |
| 2. | | | | ❐ |
| 3. | | | | ❐ |
| 4. | | | | ❐ |
| 5. | | | | ❐ |

| Next Meeting Date: | Meeting Time: | Meeting Place: |
|---|---|---|
| | | |

| Title: | | |
|---|---|---|
| Duration: | Meeting Date: | Meeting Time: |
| Location: | Called By: | Type: |
| Facilitator: | Note Taker: | Time Keeper: |
| Attendees: | | |

| Agenda | |
|---|---|
| Topic: | Presenter: |
| 1. | |
| 2. | |
| 3. | |
| 4. | |
| 5. | |

| | Action | Person Responsible | Deadline | |
|---|---|---|---|---|
| 1. | | | | ❒ |
| 2. | | | | ❒ |
| 3. | | | | ❒ |
| 4. | | | | ❒ |
| 5. | | | | ❒ |
| 6. | | | | ❒ |
| 7. | | | | ❒ |
| 8. | | | | ❒ |
| 9. | | | | ❒ |
| 10. | | | | ❒ |

Notes:

## Agenda

| Topic: | Presenter: |
|---|---|
| 1. | |
| 2. | |
| 3. | |
| 4. | |
| 5. | |

| | Action | Person Responsible | Deadline | |
|---|---|---|---|---|
| 1. | | | | ❒ |
| 2. | | | | ❒ |
| 3. | | | | ❒ |
| 4. | | | | ❒ |
| 5. | | | | ❒ |

## Agenda

| Topic: | Presenter: |
|---|---|
| 1. | |
| 2. | |
| 3. | |
| 4. | |
| 5. | |

| | Action | Person Responsible | Deadline | |
|---|---|---|---|---|
| 1. | | | | ❒ |
| 2. | | | | ❒ |
| 3. | | | | ❒ |
| 4. | | | | ❒ |
| 5. | | | | ❒ |

| Next Meeting Date: | Meeting Time: | Meeting Place: |
|---|---|---|
| | | |

| **Title:** | | |
|---|---|---|
| **Duration:** | **Meeting Date:** | **Meeting Time:** |
| **Location:** | **Called By:** | **Type:** |
| **Facilitator:** | **Note Taker:** | **Time Keeper:** |
| **Attendees:** | | |

| **Agenda** | |
|---|---|
| **Topic:** | **Presenter:** |
| 1. | |
| 2. | |
| 3. | |
| 4. | |
| 5. | |

| | **Action** | **Person Responsible** | **Deadline** | |
|---|---|---|---|---|
| 1. | | | | ❐ |
| 2. | | | | ❐ |
| 3. | | | | ❐ |
| 4. | | | | ❐ |
| 5. | | | | ❐ |
| 6. | | | | ❐ |
| 7. | | | | ❐ |
| 8. | | | | ❐ |
| 9. | | | | ❐ |
| 10. | | | | ❐ |

**Notes:**

## Agenda

| Topic: | Presenter: |
|---|---|
| 1. | |
| 2. | |
| 3. | |
| 4. | |
| 5. | |

| | Action | Person Responsible | Deadline | |
|---|---|---|---|---|
| 1. | | | | ❐ |
| 2. | | | | ❐ |
| 3. | | | | ❐ |
| 4. | | | | ❐ |
| 5. | | | | ❐ |

## Agenda

| Topic: | Presenter: |
|---|---|
| 1. | |
| 2. | |
| 3. | |
| 4. | |
| 5. | |

| | Action | Person Responsible | Deadline | |
|---|---|---|---|---|
| 1. | | | | ❐ |
| 2. | | | | ❐ |
| 3. | | | | ❐ |
| 4. | | | | ❐ |
| 5. | | | | ❐ |

| Next Meeting Date: | Meeting Time: | Meeting Place: |
|---|---|---|
| | | |

| Title: | | |
|---|---|---|
| Duration: | Meeting Date: | Meeting Time: |
| Location: | Called By: | Type: |
| Facilitator: | Note Taker: | Time Keeper: |
| Attendees: | | |

## Agenda

| Topic: | Presenter: |
|---|---|
| 1. | |
| 2. | |
| 3. | |
| 4. | |
| 5. | |

| | Action | Person Responsible | Deadline | |
|---|---|---|---|---|
| 1. | | | | ❒ |
| 2. | | | | ❒ |
| 3. | | | | ❒ |
| 4. | | | | ❒ |
| 5. | | | | ❒ |
| 6. | | | | ❒ |
| 7. | | | | ❒ |
| 8. | | | | ❒ |
| 9. | | | | ❒ |
| 10. | | | | ❒ |

Notes:

| Agenda | |
|---|---|
| **Topic:** | **Presenter:** |
| 1. | |
| 2. | |
| 3. | |
| 4. | |
| 5. | |

| | Action | Person Responsible | Deadline | |
|---|---|---|---|---|
| 1. | | | | ❐ |
| 2. | | | | ❐ |
| 3. | | | | ❐ |
| 4. | | | | ❐ |
| 5. | | | | ❐ |

| Agenda | |
|---|---|
| **Topic:** | **Presenter:** |
| 1. | |
| 2. | |
| 3. | |
| 4. | |
| 5. | |

| | Action | Person Responsible | Deadline | |
|---|---|---|---|---|
| 1. | | | | ❐ |
| 2. | | | | ❐ |
| 3. | | | | ❐ |
| 4. | | | | ❐ |
| 5. | | | | ❐ |

| Next Meeting Date: | Meeting Time: | Meeting Place: |
|---|---|---|
| | | |